Hybrid models for hydrological forecasting:
Integration of data-driven and conceptual modelling techniques

DISSERTATION

Submitted in fulfilment of the requirements of
the Board for Doctorates of Delft University of Technology
and of the Academic Board of the UNESCO-IHE
Institute for Water Education
for the Degree of DOCTOR
to be defended in public
on Friday 4th of September 2009 at 15:00 hours
in Delft, The Netherlands

by

Gerald Augusto CORZO PEREZ

born in Cúcuta, Colombia.
Master of Science in Hydroinformatics
UNESCO-IHE Delft, the Netherlands

CRC Press
Taylor & Francis Group
Boca Raton London New York

CRC Press is an imprint of the
Taylor & Francis Group, an **informa** business

A BALKEMA BOOK

This dissertation has been approved by the supervisor:

Prof. dr. D.P. Solomatine

Members of the Awarding Committee:

Chairman	Rector Magnificus, TU Delft
Prof. dr. A. Mynett	Vice-Chairman, UNESCO-IHE
Prof. dr. D. P. Solomatine	TU Delft/UNESCO-IHE (supervisor)
Prof. dr. R. K. Price	TU Delft/UNESCO-IHE
Prof. dr. S. Uhlenbrook	UNESCO-IHE/VU Amsterdam
Prof. dr. H. H. G. Savenije	TU Delft
Prof. dr. D. Rosbjerg	Technical University of Denmark
Dr. M. Werner	UNESCO-IHE, Deltares
Prof. dr. N. van de Giesen	TU Delft (reserve)

CRC Press
Taylor & Francis Group
6000 Broken Sound Parkway NW, Suite 300
Boca Raton, FL 33487-2742

First issued in hardback 2017

CRC Press/Balkema is an imprint of the Taylor & Francis Group, an informa
business

No claim to original U.S. Government works

ISBN 978-0-415-56597-4 (pbk)
ISBN 978-1-138-43400-4 (hbk)

Published by:
CRC Press/Balkema
PO Box 447, 2300 AK Leiden, The Netherlands
e-mail: Pub.NL@taylorandfrancis.com
www.crcpress.com www.taylorandfrancis.co.uk www.balkema.nl

Visit the Taylor & Francis Web site at
http://www.taylorandfrancis.com

and the CRC Press Web site at
http://www.crcpress.com

This thesis is dedicated to my mother Maria, father Fabio and my Daughter Geraldine, for their endless love, encouragement and support

SUMMARY

Operational hydrological forecasting is based on extensive use of various types of hydrological models. The most popular ones are conceptual models, followed by more detailed process distributed models. Empirical (statistical) models are used as well, and in the last decade they received more attention due to the appearance of data-driven models which, in essence, are empirical models that use the methods of machine learning (computational intelligence). A wide choice of models presents a certain challenge for a practitioner, who will have to select and integrate adequate models, and link them with the data sources. Recently a number of studies addressed the problem of integrating different modelling paradigms, and it has been shown that this approach leads to an increased accuracy of forecasts, and that more studies are needed to develop a consistent modelling framework and to test it in various situations. In this research various ways of integrating models for simulation and forecast are explored.

The increasing number of extreme and unexpected flood situations in recent decades has led to a growing interest to more accurate flood forecasting systems. These systems are necessary to provide warning against flooding preventing loss of life and minimizing damage to both properties and livestock. On the other hand, low flow forecasts are also important in the fields of water supply management, industrial use of freshwater, optimization of reservoir operations, navigation and other water-related issues. The aim of modellers is to increase model accuracy, and extend the forecast lead time. Better weather forecasts and more accurate data play here the leading role, but model improvements and the integration of different models have a lot of potential as well.

The objectives of representing a hydrological phenomenon by a model and the data availability determine the choice of modelling paradigm. In general, models used for streamflow forecasting can be grouped into three classes: a) physically-based (PBM) (often distributed) models based on a detailed representation of the processes; b) conceptual models and their more sophisticated version called process-based models (PRBM), including the so-called "semi-distributed" versions, which are important to improve process basis of predictions (land use and climate change scenarios); and c) empirical statistical or data-driven models (DDM) based on historical data about the modelled processes. PBMs are more commonly used for the interpretation of processes in river basins. These models involve a high number of physical parameters that are determined on the basis of expert knowledge, field analysis and/or in complex situation by automated calibration techniques. They are often used in assessing flood situations where the information of the expert is combined with the capacity of detailed models.

Often, however, there is not enough data to build PBMs, and for operational flow forecasting the detailed representation of a basin is not necessary. Therefore, conceptualized methods like PRBMs and data oriented techniques like DDMs are often preferred alternatives in real-time operational flow forecasting systems. The PRBM and DDM have a different basis: A PRBM's structure is based on simplified descriptions of the physical processes, whereas a DDM normally represents the mapping from the set of input variables to the output. It is commonly argued that the features of the PRBMs are missing in the DDMs and vice versa. Due to the differences in these two paradigms integrating such models is a challenging task.

Both DDMs and PRBMs are widely accepted and researched, and they have properties useful for different types of problems. When making a decision on which type of model is the most appropriate for a particular purpose, one has to consider the possibility of integrating both modelling approaches. Often models that combine different paradigms are called *"hybrid"*. In such a hybrid approach the best features of both approaches should be preserved: physical concepts of hydrological science in the PRBM and the power of encapsulating the historical data in the DDM. In hybrid modelling the different sub-models are typically responsible for modelling particular sub-processes, so partitioning of the input space using different physical concepts and/or mathematical constructs, and the subsequent integration of model outputs is needed.

As a step forward in flow simulation and forecasting this dissertation explores the use of integrated solutions with process-based and data-driven models. For this purpose it is proposed to use a hybrid modelling framework, and base it on the "principle of modular modelling".

The *main objective* of this research is to investigate the possibilities and different architectures of integrating hydrological knowledge and models with data-driven models for the purpose of operational hydrological forecasting, and to test them on different case studies. The models resulting from such integration are referred to as *hybrid models*. The following specific objectives were formulated:

1. Explore the various architectures and develop a framework for *hybrid modelling* combining *data-driven* and *process-based hydrological* models in operational hydrological forecasting, especially in the flooding context.

2. Further explore, improve and test the principle of *modular modelling* allowing for building data-driven and hybrid models.

3. Further explore, improve and test the procedures *optimizing the structure* of data-driven models, including those that work as complementary and error correction models.

4. Evaluate the applicability of modular modelling schemes in other related problems, like downscaling weather information for hydrological forecasting.

This research introduces and develops hybrid modelling principles based on modular models. A general classification of hybrid models and logical framework for hybrid modelling are developed. On the basis of the framework, modular model concepts are developed and tested on a number of case studies.

Three main principles of modularization of models considered are: spatial, temporal and processes-based. The main case study for the spatial analysis is the Meuse river basin. The Dutch Ministry of Public Works (Rijkswaterstaat) uses the hydrological modelling system "Hydrologiska Byråns Vattenbalansavdelning" (IHMS-HBV). It represents 15 sub-basins, each modelled by individual lumped conceptual model components, which are linked by a simplified routing scheme. This model is a part of the operational flood forecasting system that uses the Delft/FEWS platform developed at Deltares, and is linked to real-time feeds of the regional weather forecasts provided by the Royal Dutch Meteorological Institute (KNMI). In this research various ways of replacing some of the conceptual hydrological sub-models by local data-driven models (e.g. artificial neural networks, ANNs) are analysed. This is done on the basis of the available information (local measured discharges), and on the study of relative contribution of each sub-basin model to the overall model error. The results of such "model hybridization" show multiple advantages not only in terms of accuracy of the overall model, but also in the increase of the lead time where spatial weather information plays an important role in the simulation of low and high flow phenomena.

Experiments with temporal and process-based modular models are carried out on different types of catchments in Asia and Europe. This experiments show the advantages of combining specialized models built for different sub-processes. It is also shown that for identifying such sub-processes it is more effective to use hydrological concepts, expert judgement and knowledge, rather than the automated data analysis and clustering techniques (which however could be very useful as well). It is demonstrated how the global optimization techniques help to generate optimal model structures. Furthermore, the possibilities of using modularization in multi-step ahead forecasting are presented and compared to conventional ANN models.

An extensive sensitivity analysis of data-driven models (mainly, ANNs) is conducted in this study, along with the analysis of the dependence of different data-driven models' performance on different inputs and random initializations. These experiments confirm that flow forecasting data-driven models which use past values of discharge are dominated by autocorrelation, so that an accurate knowledge of precipitation, for a certain lead times, is less important in overall error assessments. In general, ANN models with the right choice of variables, are not so much influenced by various random initializations of weights. With the appropriate selection of variables, it appears that the correlation and average mutual information (AMI) analysis give similar results on all the cases tested in this thesis. Among all the data-driven modelling techniques tested, ANNs had the best performance. Using an ensemble of differently initialized ANNs leads to more accurate forecasts.

Parallel and complementary hybrid modelling architectures are show to improve the performance of a forecast model beyond the ANN and process-based models. Multiple combinations of ensembles and error corrector models are tested. The use of committee models (e.g. ensembles) employing ANN and the HBV models for the Meuse river basin are shown to have almost the same performance as a model with error corrector built with information from previous errors and previous states of the model. In the Meuse case study the non-linear error corrector is found to be better than the linear error correctors. The results show that adding the error corrector improves the accuracy of the HBV for the lead times which are higher than the concentration time. It appears from experiments that a single ANN cannot produce accurate forecasts for lead times higher than the characteristic lag (travel) time of the particular river. These experiments are based on the assumption of perfect rainfall forecast, but can be extended for real forecasts. In general, it is shown that the limitations of the process-based models can be overcome by complementary error correcting data-driven models.

Yet another case study relates to downscaling information from general circulation models into meteorological information at watershed scale. The modular modelling approach (based on clustering samples and building separate models for each of them) brings an improvement over conventional statistical and data-driven models. A case study in Ethiopia and data from national centre for environmental prediction (NCEP, from USA), are considered. The results show an improvement in terms of overall accuracy for precipitation, however, the results for temperature are less convincing. The latter can be explained by the fact that temperature is a more periodic variable than precipitation, and its relatively slow transition between low and high values makes it less appropriate variable for driving modular models.

In general, this research presents a hybrid modelling framework where data-driven and conceptual process-based models work in a coordinated fashion, and their role and performance are optimized. Several principles of models hybridization and modularization – spatial, temporal and processes-based – are considered and explored on a number of case studies. Advantages and disadvantages of various approaches for different lead times are evaluated and discussed. In the framework of one of the case studies, the developed models are incorporated as software components into operational hydrological forecasting system for Meuse river basin, implemented on the Delft/FEWS platform. This thesis contributes to hydrological flow forecasting and its findings, I hope, be used in building more effective flood forecasting systems.

Gerald Corzo

CONTENTS

INTRODUCTION

1.1 Background

Flood events are becoming more frequent and intense in many countries around the world. One of the major concerns in the world is the recent increase of catastrophic flood situations. Many researches point out that in the coming decades the situation may become worse due to the climate change (Palmer and Risnen, 2002). Flood management knows various approaches for controlling floods (to some extent) and for mitigating their consequences. There are structural approaches that are expensive and not always possible. Solutions like reforestation, proper urban planning and extension of flood plains, are often effective but are long term and do not always guarantee considerable reduction of flood damage. Due to the fact that in many situations it is practically impossible to prevent floods, it is important to build models and systems that are able to forecast hazardous situations with the highest possible accuracy. It is common that flood management includes flood warning system or flow forecasting system providing assessments of the spatial range and duration of flooding.

Accurate forecasting of natural phenomena with extended lead times is one of the challenges for practitioners. The river flow forecasting systems are generally supported by hydrological and river models. The requirement in terms of accuracy is always relative to the lead time required. For extended lead times, the use of weather forecast information provides the information for the conceptual models, but increases the models' uncertainty. Operational forecasting systems are not perfect, and often measurements of precipitation or water levels are missing. Complex aid models are needed for filling missing data and incorporating other available measured data through data-assimilation. All these problems make the problem of operational hydrological forecasting quite a challenging task.

The problem of flood management and improved forecasting is one of the primary application areas of hydroinformatics (Abbott, 1993; Price, 2005; Solomatine, 2005). Evolution of information and communication technologies in the last decades has lead as a consequence to an increase in the number of measurements. Therefore, nowadays, large amounts of data for various environmental variables have been collected. These are especially useful where there is limited or no domain knowledge (on physics, chemistry, and biology of the process) available. More and more hydroinformatics systems are linked in real-time to the Numerical Weather Prediction models, allowing for direct feeds of the precipitation and temperature forecasts into hydrological models.

Data availability opened up the possibilities of new modelling paradigms that have been increasingly applied in hydrological modelling in the last decades. One of such paradigms is the so-called data-driven modelling actively researched in the framework of hydroinformatics. However, the advances in computer sciences and computational intelligence, the main suppliers of technologies for this area, allow for building more accurate, optimized hybrid models incorporating different modelling paradigms in a flow forecasting system.

This study belongs to the area of hydroinformatics, and is at the interface between hydrological modelling and computational intelligence, and one of its important application areas is flood management and forecasting. The study has been conducted in the framework of the "Delft Cluster" research programme of the Dutch Government (project "Safety Against Flooding", and was possible due to the financial support of this project.

In the following sections flood forecasting systems, process-based hydrological models, and data-driven models will be characterized, and the possibilities of building on their basis hybrid models will be shown.

1.2   Flood management and forecasting

Flood management is a comprehensive area that has received a lot of attention from researchers and practitioners during the last decades. In particular, several research efforts have been supported by the EU research funds, e.g., in the MUSIC, FLOODRELIEF, FLOODsite, and a large number of other projects where serious attention was given to development of effective methods and platforms enhancing flood management. For this study it is important to position operational hydrological forecasting within the set of possible flood management measures, to characterize the operational flow forecasting process, and to identify the place of new types of computer-based models in this process.

1.2.1   Flood management measures

In river flood management two classes of measures are distinguished.

Structural measures   Construction of dams, weirs, barriers, dikes and other facilities are some of the most common structural solutions. These are not just expensive but also often do not guarantee an effective solution.

Non-structural measures These normally can be subdivided into two groups. One is the application of environmental solutions like reforestation, proper urban planning, flood plain management, etc. These solutions are long-term and expensive and do not always guarantee a reduction of flood damage, or shift it sometimes to another area. The second group is the implementation of river flood warning systems. The mitigation of flood impact in this case is based on the following premise: since the flood is impossible to prevent, it is important to have a solution to provide advice and to mitigate the possible consequences of the flood. Such a flood warning system is justified by the usefulness of the hydrological flow forecast, and reflected in anticipating the possible spatial ranges and temporal durations of floods.

In the context of river flood management, hydrological flow forecasting models are the core of warning systems, and therefore they attract the attention of managers and researchers. Advanced systems that have been deployed in the last decade include GIS visualization, possibilities to generate inundation maps, assess potential damages, connect in real-time to the various data sources and numerical weather prediction models and issue warnings across multiple communication platforms (Price, 2005; Werner, 2004; Werner et al., 2005), see Figure 1.1.

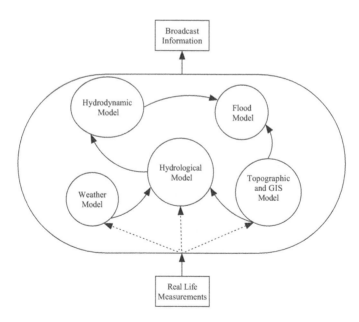

Figure 1.1: *Simplified scheme of interactions between models in a flood forecasting system*

1.2.2   Operational flow forecasting

In general, flood and flow forecasting in operational systems are only differentiated by the (water level) threshold indicating a flood situation. This flood concept is relative and therefore the scope of this research will relate more generally to "flow simulation" and "flow forecast". The concepts of flow simulation and forecasting are important for the interpretation and nature of the work presented in this thesis. A flow simulation model is defined as a model representing the hydrological processes in the basin, from the forcing variables (e.g. precipitation, temperature and others) to the river discharge. A flow forecasting model is defined as the model which receives weather forecast information as input and calculates future values of flow, typically, of river discharge.

Conventionally, flow forecasting in an operational system is performed using process-based and/or conceptual models of river basins, and not fully physically-based models. Conceptual models are generalizations of the system, that use or process the main forcing variables in an flood event. In general, soil properties, topographical information, and other complex spatial variables are managed as global and/or constant. The simplification of the physical system makes the conceptual models relatively easy to apply and allows for fast processing in an understandable manner. At the same time the models cover the general ongoing processes in the basin.

The use of more complicated spatially distributed models for multi-time-step forecasts neither allow the accuracy of the conceptual models, nor have the short processing time required for a fast response. Paradoxically, spatially distributed physically based models and energy based models often appear to be less accurate than conceptual models (Linde et al., 2007; Seibert, 1997). Diermansen (2001) presented an analysis of spatial heterogeneity on the runoff response of large and small river basins, and observed an increase in error with the increase of the level of detail in a physically based model.

This study actively uses an important class of intermediate models, the so-called semi-distributed conceptual models, as the most appropriate modelling approach for meso-scale forecasting. The studies presented by Fenicia et al. (2008) show that with simple semi-distributed flexible models, it is possible to find an appropriate hydrological distribution and regionalization of hydrological processes to better understand the river basin. Their approach allowed the characterizing the basin, having a spatial distribution of lumped sub-basins with an optimal regionalization of hydrological process.

Real-time hydrological forecasting systems are set up to work in a dynamic environment. Typical interaction of data and models in a conventional flood early warning system is presented in Figure 1.1, and Figure 1.2 presents the time line of different processes. The information from the gauges is commonly collected on a hourly or daily basis from different places in the region. Commonly this information is passed from the collecting organization database to an file transfer protocol (FTP) server. From the FTP server the information is collected and stored in a local database for further access from different data pre-processing and modelling software. This time T_1 (Figure 1.2), is higher when the system depends on the collection of measurement of many places (e.g. short number of hours and could reach 1 day in extreme cases). Measured data and weather forecast information are prepared. A number of pre-processes are run in order to fill missing data, to interpolate regional information and to validate its values (T_2).

Weather forecasting model results come in intervals of 6, 12 or more hours; intermediate values are calculated by weighting schemes or interpolation (T_3). Ensemble weather forecasts are commonly supplied in a range of equally probable alternative rainfall vectors (e.g. 51 ensemble members, European Centre for Medium-Range Weather Forecasts, ECWMF). Each ensemble of weather information is produced for each grid point over the river basin. This implies that large databases are required to manage all the weather forecast information. The gauge measured data is used to update the model states in such a way that the starting forecast should be perfect (T_4). The simulation of hydrological models, hydrodynamic models and error corrector models is done by iteration, step by step, till the forecast horizon is reached (T_5). When all the simulations are finished, a forecaster follows a procedure to generate a public report or an active alarm, if required (T_6).

The time available to issue the warning depends on all these procedures. The highest delays are typically T_1, T_5 and T_6, since T_1 depends on a good communication and management policy. T_5 is the time used by the model, and T_6 is the time needed to issue the documents and broadcast the information. In the case of the Meuse flow forecasting system considered in this thesis, 52 possible solutions of the model from 3 agencies are used to simulate scenarios for a model that has 15 sub-basin models. The model (including several connected conceptual hydrological models and a hydrodynamic river model) is run several times in a day on a cluster of computers and the computational time is relatively small. (Note that in real-time situations there could be delays in producing and communicating the forecasts and warnings due to computational and administrative barriers).

Weather and hydrological models form the basis of the flood forecast. These models have to interact, so that one feeds the other. All models and data sources bring uncertainties, and the study of these uncertainties and their propagation through the model chain is nowadays an important issue (Candela et al., 2003; Glemser and Klein, 2000). Commonly it is agued that the main source or error and uncertainty is the weather forecast model (quantitative precipitation forecast), and this prompts additional efforts aimed at capturing part of the associated dynamics (Bartholmes and Todini, 2005; Tu et al., 2004).

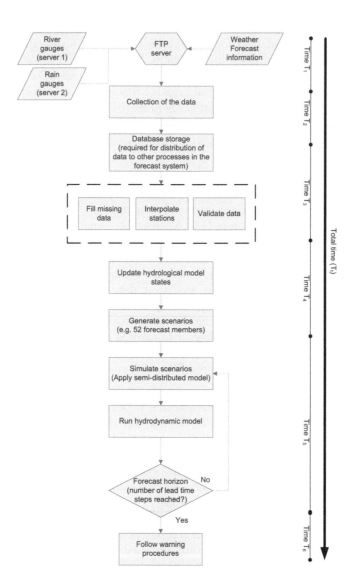

Figure 1.2: *Chart of operational flow forecasting processes*

1.3 Hydrological models

1.3.1 Classification

Hydrological models have been distinguished and classified on the basis of their function and objectives, their structure, and their level of spatial disaggrega-

tion. Since there are various ways to classify hydrological models, here it is
introduced the classification (more-or-less standard one) that is used in this
thesis.

> *Physically-based models (PBM)*: Are generic equations for flow at a point
> with the model space discretized in accordance with the numerical so-
> lution of the equation used (Price, 2009). In general terms, this models
> represent the natural system using the basic mathematical representation
> of the flow at a point; based on the conservation of mass, momentum and
> energy. For river basin model representations, a physically-based model
> in practice has to be also fully distributed (Refsgaard, 1997).
>
> *Conceptual models*: Physically-base equations relating output to input
> for the model discretized according to the identification of physical boun-
> daries (Price, 2009). In general terms the concept have been described
> in the past as models rising from simple verbal descriptions to equations
> governing relationships or 'natural laws' that purport to describe rea-
> lity (Refsgaard, 1997). However, nowadays this is refereed to a more
> comprehensive type of models that attempt to simulate to a greater or
> lesser extent, the most important perceived hydrological mechanisms of
> the catchment response to rainfall, e.g. interception, evapotranspiration,
> infiltration, and both groundwater and surface water flow routing, etc.,
> using prescribed physical plausible empirical and heuristic mathemati-
> cal relations. Although not "physically-based", in the sense of using
> pure physically-based equations, they are nonetheless "physically inspi-
> red" (O'Connor, 2005).
>
> *Process based models (PRBM)* are a relative new way to refer to con-
> ceptual models that have a clear process structure (example: TAC and
> TAC^D, Uhlenbrook et al., 2004). This is an intermediate model which oc-
> cupies a position between the distributed physically-based model and the
> hydrological lumped model. They are in contrast to the physically-based
> models that are fully distributed and take account of spatial variations in
> all variables and parameters. A PRBM can be a semi-distributed model
> that takes into account spatial variation but as a integration of several
> lumped small catchments.
>
> *Data-driven models (DDM)*: empirical models based on learning from
> data, and associated with machine learning (computational intelligence)
> algorithms. They are sometimes referred to as models that induce causal
> relationships between sets of input and output time series data in the
> form of a mathematical device, which in general is not related to the
> physics of the real world (Solomatine and Price, 2004).

The most common way to represent the hydrology of a region, in terms of a
conceptual model, is through the use of a storage-based simulation. The stor-
ages (buckets) represent surface water storage and groundwater components

and each component has one or more coefficients that are calibrated with a fixed time step. There is no difference in time between the input(s) and output(s) of these models (however, it can be also said that such a model makes a one-step ahead forecast). The output at each time step is calculated mainly with the state variables of the model. This memory or state variables include the soil moisture states, and groundwater levels and others, at the previous time step.

1.3.2   HBV process-based model

The IHMS-HBV model, extensively used in this work, is a semi-distributed conceptual rainfall-runoff model originally developed in a software product by the Swedish Meteorological and Hydrological Institute (SMHI, Bergström and Forsman, 1973, ,(Hydrological Bureau Waterbalance-section)). It is considered to be semi-distributed due to the fact that a basin may be separated into a number of sub-basins and each one of these is categorized according to elevation and vegetation. The soil moisture of the HBV modelling tool is based on a modification of the bucket theory in that it assumes a statistical distribution of the storage capacities in a basin. This assumption makes the tool independent of scale as long as this distribution function is stable Lindström et al. (1997). HBV is a process based model with conservation of mass and a general formulation for evapotranspiration, snow component, soil moisture and other important hydrological processes.

HBV requires input data such as precipitation (on daily or shorter time steps), daily or shorter air temperature (if snow is present), daily or monthly estimates of evapotranspiration, daily runoff records for calibration and validation, and geographical information about the river basin. The principal components of the model are precipitation, evapotranspiration, storage and runoff. These components are related to each other in a given period of time through the water balance equation given as:

$$P - E_A - \Delta S/\Delta t = Q \qquad (1.1)$$

where:
P =precipitation (mm/day),
E_A =actual evapotranspiration (mm/day),
Q =runoff (mm/day),
ΔS=change in basin storage (mm), per time step Δt (day)
Note: the equation holds true as long as no water passes the system boundaries (e.g. groundwater flows from other basins).

The model has gradually been developed into a semi-distributed model. Distribution of inputs in the model is guaranteed through the use of subbasins (considered as primary hydrological similar units) in the schematisation. Further distribution within a subbasin is possible in terms of the area-elevation distribution and a crude classification of land use into forest, open area and lake (Bergström and Forsman, 1973).

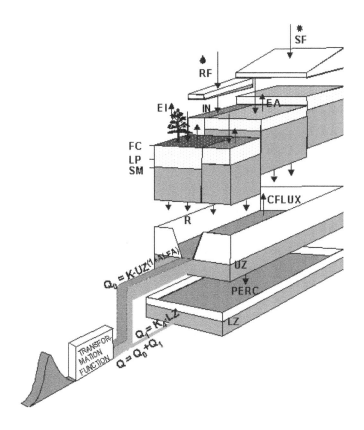

Figure 1.3: *HBV model schematization*

The basic routines to be characterized cover snow accumulation and snow melt routine, soil moisture accounting routine, runoff generating routine and routing procedure Lindström et al. (1997).

Snow melt routine The standard snowmelt routine of the HBV model is a degree-day approach, based on air temperature, with a water holding capacity of snow which delays runoff. Melt is further distributed according to the temperature lapse rate and is modelled differently in forests and open areas. A threshold temperature, TT, is used to distinguish rainfall from snowfall.

Soil Moisture Accounting Routine. This routine is the main part controlling the formation of runoff. It is principally the simulation of the water balance equation. The actual evapotranspiration is computed as a function of the soil moisture conditions and the potential evapotranspiration (PET[mm/day]). When the soil moisture exceeds the storage threshold (LP[-]), water evaporates at the

Table 1.1: *Main variables of the HBV model*

Variable	Description
SF	Zone snowfall $[\text{mm}/\text{t}^a]$
RF	Zone rainfall $[\text{mm}/\text{t}]$
R	Recharge $[\text{mm}/\text{t}]$
EA	Actual evapotranspiration $[\text{mm}/\text{t}]$
EI	Interception evaporation $[\text{mm}/\text{t}]$
IN	Infiltration to soil $[\text{mm}/\text{t}]$
FC	Maximum soil moisture content [mm]
LP	Limit for potential evapotranspiration [-]
SM	Soil Moisture [mm]
CFLUX	Maximum capilarity flow $[\text{mm}/\text{t}]$
UZ	Storage in upper response box [mm]
PERC	Maximum percolation rate $[\text{mm}/\text{t}]$
LZ	Storage in lower response box [mm]
ALPHA	Response box parameter [-]
K	Recession coefficient $[t^{-1}]$
K_1	Recession coefficient $[t^{-1}]$
K_4	Recession coefficient $[t^{-1}]$
Q_0	Outflow from upper response box $[\text{mm}/\text{t}]$
Q_1	Outflow from lower response box $[\text{mm}/\text{t}]$
Q	Outflow from transformation function $[\text{mm}/\text{t}]$

[a]t is a time unit, can be defined in hours or days

potential rate. At lower soil moisture values a linear relation between the ratio AET/PET and soil moisture is used. Three parameters namely, β , LP, and FC[mm] control the hydrological processes in this routine as shown in Equation 1.2 β controls the contribution to the response function $\left(\frac{\Delta Q}{\Delta P}\right)$ for each millimetre of rainfall or snowmelt.

$$\frac{R}{IN} = \left(\frac{SM}{FC}\right)^{\beta} \tag{1.2}$$

- LP is the soil moisture value above which evapotranspiration reaches its potential value, usually given as a ratio.

- FC [mm] is the maximum soil moisture storage (Field capacity) in the model. FC is a model parameter and not necessarily equal to measured values of 'field capacity'.

- IN[mm/t] is the maximum soil moisture storage in the model.

- R[mm/t] is the recharge.

- SM [mm] is the soil moisture storage in the model .

The Runoff Generation and Routing Routines. Once the water balance is establis-hed for the snow and soil moisture accounting routines the response function transforms the excess water into runoff and then into a hydrograph. This func-tion consists of one upper non-linear reservoir, and one lower linear reservoir and one transformation function. These reservoirs are the origin of the quick and slow components of the hydrograph, respectively. The response function is governed by five empirical parameters:

- K_1, K_4 and α which are recession parameters

- PERC the percolation capacity of the soil

- MAXBAS which is the parameter of the transformation function which

represents the time base of the resulting hydrograph. The various hydro-logical processes and the equations governing these processes are presented in Figure 1.3.

Procedures for model calibration. Model calibration (parameter estimation) in-volves the automatic and/or manual adjustment of model parameters to mi-nimize the difference between observed and simulated values. The assessment of the goodness of fit can be carried out using either subjective or objective methods. In this study we also used external calibration tools based on more sophisticated randomized search methods (e.g. Solomatine, 1999).

Performance criteria. Uncertain inputs, model structure and initial conditions are inherent ingredients in modelling the hydrology of a region (Leavesley et al., 2002). It is therefore a common practice to develop model performance or validation criteria in order to test the integrity of the modelling exercise. To assess the performance of the model for each test a standard set of criteria of calibration and validation is normally used. The following are some of the most widely used performance measures.

- The Coefficient of Efficiency, (COE, Nash and Sutcliffe, 1970, Equation 2.4)

- Joint plots of the simulated and observed hydrographs

- Normalized Root Mean Square Error (NRMSE, Equation 2.3)

For this thesis, these error measures and others are explained in chapter 2.

1.4   Data-driven models

Data driven models are models based on computational intelligence algorithms that are typically associated with learning from data. They are sometimes referred to as models that induce causal relationships or patterns between sets of input and output time series data in the form of a mathematical device, which in general is not related to the physics of the real world simulation (Solomatine and Price, 2004).

Application of computational intelligence algorithms, especially artificial neural networks, to model hydrological behaviour has been actively explored in recent years (ASCE, 2000a; See and Openshaw, 1999; Solomatine and Dulal, 2003). Most of the results show that such models often outperform in terms of accuracy other conventional modelling techniques (Brath et al., 2002; Toth and Brath, 2002; Toth et al., 2000). The well known problem of data driven models with respect to extrapolation, educed in the training stage, seems to be solved by the use of an additional physically based or conceptual model that is, run on a number of events with a high return period and then the results are used to train a DDM (Hettiarachchi et al., 2005). Additionally, data-driven approaches have shown to be improved by using model combinations: techniques like mixture of models and committee machines have open new modelling alternatives to solve highly complex problems. Therefore, they should be seen as an important alternative to be considered in forecasting hydrological flows, at one or multiple time steps. However, many practitioners in operational flow forecasting still have reservations about data-driven models and are more comfortable with the more traditional conceptual models.

Ideas of integrating various types of models are becoming more and more popular among researchers, and have gradually become known to practitioners as well. This new area is promising because the use of computational intelligent algorithms has been shown to extend the modelling capacities of conventional models. It is one of the main objectives of this research to show the different options and results on the use of these two methodologies in one single modelling process.

Hybrid models

Exploration of the use of data-driven models (statistical, and those using the methods of computational intelligence) in forecasting environmental variables provides evidence that, for many problems, they could be accurate estimators. However, the knowledge representation in this type of model is not explicit, and therefore normally not useful for obtaining information about the ongoing processes for critical situations. They are based on the analysis of the mathematical relationships between the variables describing the system, whose behaviour is to be predicted. The data-driven models are also highly dependent on the available data, and are commonly referred to as grey or black box models. In important characteristic is that their accuracy results can be used as reference for hydrological problems (Lischeid and Uhlenbrook, 2003).

Conceptual hydrological and DDM modelling paradigms seem to have complementary features for their joint use in hydrological modelling. However, a general framework needs to be formulated. In this sense, a number of questions are to be answered. What are the possible ways to integrate them? What is the performance of an integrated (hybrid) model? What are the advantages of integrated (hybrid) modelling approach?. The integration of these modelling techniques is not straightforward, and a way forward is undertaken in this thesis.

The main subject of this thesis relates to the ways of integrating data-driven models with hydrological knowledge and hydrological models for flow forecasting. Hybrid modelling is a relatively new concept that emerges from existing modelling techniques. For this purpose, hydrological flow modelling, data-driven models and modular models, are reviewed and conceptualized. A simplified mind map of models and ideas may help in such conceptualization (Figure 1.4)

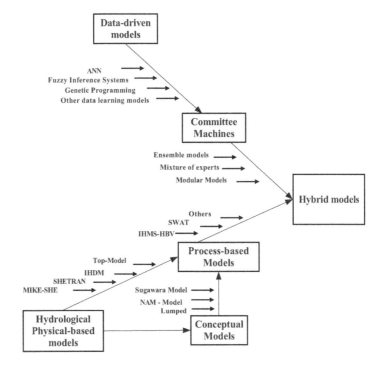

Figure 1.4: *Two branches, data-driven and physical-based modelling, leading to hybrid models*

1.5   Objectives of the research

The main objective of this research is to investigate the possibilities and different architectures of integrating hydrological knowledge and models with data-driven models for the purpose of operational hydrological forecasting, and to test them in different case studies. The models resulting from such integration are referred to as *hybrid models*.

The specific objectives can be summarized as follows:

- Explore the various architectures and develop the framework of *hybrid modelling* combining *data-driven* and *process-based hydrological* models in operational hydrological forecasting, especially in the flooding context.

- Further explore, improve and test the principle of *modular modelling* allowing for building data-driven and hybrid models.

- Further explore, improve and test procedures for *optimizing the structure* of data-driven models, including those that work as complementary and error correction models.

- Evaluate the applicability of modular modelling schemes in other related problems, like downscaling weather information for hydrological forecasting.

Key research questions

To reach the objectives, different research questions have to be formulated. The main research question is formulated as follows.

How can the hybrid modelling approach be used in hydrological forecasting, and what are the modelling architectures to be used for this purpose?

- How can hybrid and modular modelling architectures be classified?

- Are there gains in performance when hybrid models are used in typical hydrological modelling tasks?

- What are the advantages of including data-driven models in a semi-distributed process-based model of the large river basin (on an example of the Meuse basin) in the context of operational forecasting?

- What are the advantages of particular architectures and the optimization of data-driven, process-based, and hybrid and modular models (including data-assimilation with error correctors and ensembles)?

- Can the modular modelling approach help in the statistical downscaling of weather information into predicted precipitation values for use in hydrological models at the basin scale?

1.6 Terminology

The terminology related to hydrological modelling and forecasting is relatively well established, but one may also observe certain changes in terminology over time, and certain preferences of various authors and schools. In computational intelligence the terminology is perhaps less established, and there are many similar methods that are called differently by different authors.

Committee machines (CM): This term has close meaning to modular models, and the two are often interchangeable. A committee machine is a term originally linked to the use of artificial neural networks. The concept of committee machines is not new and can be related to the work by Nilsson (1965); the artificial neural network structure considered by him consisted of a layer of elementary perceptrons followed by a vote-taking perceptron in the second layer (Haykin, 1999). In the context of data-driven modelling, it is possible to classify the CMs (Price et al., 1996; Solomatine, 2005) with respect to the way the splitting is performed and how subsequently trained models are used with new input data: only one of them, so-called model mixtures, statistically-driven; or combination of all of them according to a ensemble averaging scheme that may involve fuzzy logic.

Delft-FEWS: refers to a general integration tool for flood forecasting developed over several years at Delft Hydraulics in the Netherlands (now: Deltares). The main philosophy underlying the software system is to provide an open architecture, that permits the integration of arbitrary hydrological and river routing models with meteorological data and numerical weather forecasts. In its actual form Delft-FEWS constitutes a collection of platform-independent software modules, linked to a central database. In this study the operational hydrological forecasting system for the Meuse river basin based on the HBV model and incorporated into Delft-FEWS was used.

Hydrological forecasting: This is the estimation, or calculation in advance, of flow conditions based on the analysis of data and the use of models. A hydrological conceptual model generates a forecast value that represent the predicted value of the river discharge in the time step used in the calibration of the model.

Hybrid models: these are composed of models originating from different paradigms or sciences. In this research we will refer to the integration of computational intelligence (data-driven) and hydrological (physically-based) sciences.

IHMS-HBV or HBV: The Integrated Hydrological Modelling System, is a semi-distributed conceptual rainfall-runoff model originally developed by the Swedish Meteorological and Hydrological Institute (SMHI, or IHMS). There are also other implementations of HBV.

Modular models (MM): A modular model is a model with a structured representation of information in a particular domain. A modular model includes the definitions of modules and links between them, the rules used to build such model, and how to use it. In the context of data-driven modelling, a DDM (e.g., a neural network) is said to be modular if the computation performed by the network can be composed into two or more modules (subsystems) that operate on distinct inputs without communicating with each other. The outputs of the models are mediated by an integrated unit that is not permitted to feed information back to the modules. In particular, the integrating unit both decides how the outputs of the models should be combined to form the final output, and determines what modules should learn which training patterns (Osherson et al., 1990).

Physically-based models (PBM): Are generic equations for flow at a point with the model space discretized in accordance with the numerical solution of the equation used (Price, 2009).

Simulation: This is the imitation of some real thing, state of affairs, or process. The act of simulating something generally entails representing certain key characteristics or behaviours of a selected physical or abstract system.

Training or Learning: A computer model or program is said to learn from experiences E with respect to some class of task T and performance measure P, if its performance at tasks in T, as measured by P, improves the experience E (Mitchell, 1997). It is common to speak about training a data-driven model on some past measured data (representing experience, Mitchell, 1998).

1.7 Outline

The thesis is outlined as follows (Figure 1.5):

- Chapter 1 introduces the context of the research, background and objectives.

- Chapter 2 introduces a classification of hybrid models. This chapter also covers the description of the different performance measures used in this thesis.

- Chapter 3 covers the hybrid modelling methodology. These are presented in three main schemes that are applied to different basins in subsequent chapters.

- Chapter 4 describes the basic principles and existing problems of defining the optimal structure and using the data-driven modelling techniques. Different modelling algorithms are introduced and the procedures used in chapter 5, 6, 7 and 8 are explained. The results of applying these techniques on a case study are compared and discussed.

- Chapter 5 covers the implementation of case studies for the modular modelling techniques presented in Chapter 3. This chapter focuses on the analysis of the performance of different modular modelling architectures presented in chapter 3, with applications to catchments in Italy, Nepal and England. This chapter ends with a discussion of the advantages and disadvantages of such methodology for operational flow forecasting.

- Chapter 6 presents an application of the developed hybrid modelling methodology to the Meuse basin hydrological forecasting. Procedures allowing for optimal spatial modularization and incorporation of data-driven models into the semi-distributed IHMS-HBV model are presented.

- Chapter 7 presents contrasting use of the parallel (ensembles) and serial (data assimilation) architectures of hybrid and modular models. Relationship between the forecasting horizon and the choice of modelling architecture is analysed.

- Chapter 8 introduces the application of modular models to downscaling precipitation information from measurements GCM models into precipitation values to be used in the hydrological models at basin scale. A case study in Ethiopia is considered.

- Chapter 9 presents conclusions and recommendations.

- Appendix A explores the transformation from state-space mathematical representation to an input-output mathematical formulation.

- Appendix B describes the main data-driven models algorithms used in this thesis.

- Appendix C presents results from the hourly forecast in the MEUSE using the Delft-FEWS system.

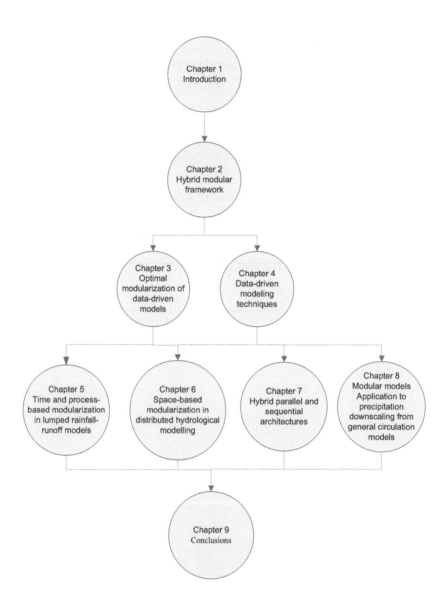

Figure 1.5: *Outline of the dissertation*

CHAPTER

TWO

FRAMEWORK FOR HYBRID MODELING

The importance of exploring hybrid modelling paradigms for flow forecasting is highlighted in Chapter 1. This chapter defines the theoretical framework for hybrid modelling and modular models. The main principles of the framework are based on the classification of the ways different modelling and knowledge paradigms are merged. The concept of modularization of models, as the most important concept for hybrid modelling, is defined and developed for its further application in subsequent chapters.

The criteria used in performance analysis of the different models explored in this thesis are presented.

2.1 Introduction

Both conceptual and data-driven models have their own advantages, disadvantages and areas of application, so it may be suggested that their combination in the form of hybrid models may bring certain gains. We cannot say that the notion of hybrid modelling in hydrology is wide spread, but it has been mentioned in several publications in recent years (Abrahart and See, 2002; Anctil and Tapé, 2004; Corzo and Solomatine, 2006a; de Vos and Rientjes, 2005; Solomatine and Price, 2004). Recent and relatively old studies attempt to combine different modelling paradigms in hydrology.

A data-driven model is built on past data (measurements), and the structure of the model is fitted through training (calibration). On the other hand, conceptual and process-based models are based on a consideration of hydrological processes and include generalized parameters that assume a physical simplification of the overall hydrological system. When these two paradigms are combined one should think of fitting together a number of concepts and variables that may relate to different areas of science; so a number of assumptions and simplifications need to be made.

19

Modelling the notion of a model state (reflecting the states of nature) plays an important role in the development process. A conceptual tank model can be mathematically represented using state-space mathematical representations (Singh and Frevert, 2002). This is explored in appendix A). On the other hand, data-driven models rarely have interpretable states, and work mainly as input-output models. Such a concept is useful when analyzing the transformation from one state to other. These formulations are important for understanding the principles of the modelling process.

Even when several models belonging to one paradigm are combined, for example, in an ensemble, there could be methodological problems to resolve. This concerns the notion of model state, which, when several models with the same states are combined, becomes undefined since the multiple model states do not reflect the states of nature any more. When models of different paradigms are combined in a hybrid model, the situation is even more confusing. A modeller simply has to live with this "deviation from the theory", being compensated by the fact that the resulting model may become more accurate.

This chapter reviews the literature relating to hybrid modelling, explores the ways in which hybrid models could be built, and suggests a possible classification of such models.

2.2   General considerations and assumptions

A hybrid model is a relatively new concept in hydrological forecasting. Therefore, it would be right to try to classify the approaches for integrating the different models. Classification of hybrid models can be based on a number of criteria. In this work we have chosen to take into account the following considerations and assumptions:

- The amount of domain (hydrological) knowledge used to build the model. The amount or degree of knowledge representation is an abstraction that needs to be defined. There is a spectrum of possibilities, and for the purpose of this thesis it is assumed that in the knowledge representation there are two extremes (data-driven vs. process-based) leading to two types of models respectively. Figure 2.1 illustrates the difference in "ratio" of data and knowledge in two imaginary models, one being mainly data-driven, and the other knowledge-driven or process based.Although there is no clear measure of knowledge representation, the hypothetical measure depicted here refers to the share of knowledge related to the physical concepts over the total amount of knowledge.

- Many natural processes allow for partitioning into sub-processes which can be modelled separately. Each model will then represent a specific process, time regime or a particular geographical area (e.g. regional phenomenon). These sub-models can be of any type, or they could be a combination of models of different types.

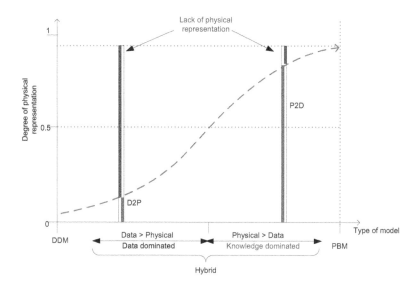

Figure 2.1: *Hybridization view of physical knowledge to data-oriented modelling approaches*

- Models of various types can be run in parallel, thus forming ensembles.

2.3   Hybrid modelling framework

2.3.1   Classification of hybrid models

Based on the possible relationships of process-based and data-driven models, it is suggested to consider the following three major classes of models (Table 2.1):

Class 1 (P2D)

This class includes data-driven models or their combinations, with the incorporated hydrological knowledge. The following presents some examples and sub-classes of this class of models.

Modular Models: This concept is based on the idea "to divide and conquer". It may be expected, that by dividing the input space into less complex and more homogenous sub-spaces, data-driven models will perform better (Osherson et al., 1990). In order to do this, physically-based concepts can be used to identify the processes, states and seasonal transitions in the data, or some

Table 2.1: *Classes of hybrid models*

Hybrid Class	Hybrid model application
Class 1(P2D): Data-driven models with the incorporated hydrological knowledge	Modular models: separation of the input-space based on hydrological knowledge (regimes, process or spatial phenomena).
	Hybrid structure: modification of data-driven model structures (use physical constrains) or identification of process and regimes from outputs (e.g. baseflow, events, others)
	Learning process: Modifying the training process of the data-driven with physical constrains.
Class 2 (D2P): Process models using data-driven techniques, or with some components replaced by DDMs.	Component model: Use data driven models to solve complex processes in a physically-based model.
	Knowledge discovery: data classifiers to group patterns of physical processes.
Class 3 (DPPS): Use of data-driven models in parallel (e.g., ensembles) or sequentially (e.g. data assimilation)	Ensemble: combine the results of both types of models
	Data assimilation (series): Use the data-driven models to identify systematic errors and correct them.

partitioning (clustering) techniques can be employed. A number of studies in this area have been conducted.

- One of the first examples of using modular models in hydrological forecasting was done by See and Openshaw (1999, 2000). Their methodology was based on an input data split as a pre-process for neural network models, and then they were integrated by a set of fuzzy rules. The input used a self-organizing neural network to create a number of clusters. Such cluster, were visually interpreted as parts of the hydrograph. Finally, output rules were used to combine those clusters. A genetic algorithm was used to find the best clusters. The potential of using this type of model is highlighted for real time flow forecasting. However, the classification performed by the self organizing map may not be adequate for low flow phenomena.

- A semi-blind separation based on wavelet analysis was used by Wang and Ding (2003). This hybrid approach was tested on short and long term time series of daily discharge and ground water level data respectively.

Their results suggest that this approach could increase the forecast accuracy and prolong the lead time of the prediction. However, their input separation results are hardly interpretable. The performance measure used was the percentage of absolute errors falling in a certain region of the time series.

- Wang et al. (2006) built an autoregressive model of flow forecasting and used the threshold separation of samples based on empirical formulations. In this work they also present a cluster based and a periodic separation. The best performance on multiple lead time steps was obtained by the periodic neural network (based on fuzzy c-means clustering). However, their analysis lack of validation and they did not include precipitation data, which is the main forcing of extreme events.

- Jain and Kumar (2007) performed hydrological time series forecasting based on de-trending and de-seasonalising of time series; they used the term hybrid neural network.

Many of the modelling approaches to build modular models mentioned above did show improvements in model performance. Most of these methods however, use automated methods to identify the sub-processes (regimes) and typically do not explicitly use hydrological knowledge in the modularization process.

There are examples of explicit use of hydrologic knowledge in building modular models:

- A modular model architecture was presented by Zhang and Govindaraju (2000a,b), who explored the creation of rules for low, medium and high flow conditions and the use of a soft classification method. This approach was applied to a monthly prediction of discharge events. The approach was tested on 180 samples only, and on this small data set the advantages of using modular neural networks compared with an overall singular neural network were marginal.

- Solomatine and Xue (2004) presented an approach to building modular rainfall-runoff models where, based on expert judgement encapsulated in simple rules, input data was partitioned into several subsets, and separate ANN or M5 model tree models were built for each subset. Building separate DDMs for various types of hydrometeorological conditions resulted in an increased accuracy of the forecasts.

- Solomatine and Siek (2006) developed a modification of the M5 model tree algorithm (M5flex) allowing an expert to control the process of building modular piece-wise linear regression models.

- Corzo and Solomatine (2007a,b) explored the use of empirical formulations optimized for real time forecasting, which were compared with non-empirical (automatic data clustering) methods. The former showed better accuracy and interpretability, although in multiple-time-steps forecast both of the techniques (empirical and non-empirical) were accurate. This approach is further explored further in this thesis by applying it to hourly and daily hydrological rainfall runoff models.

Hybrid structure of a data-driven model: The representation of a data-driven model can be modified in such a way that some of the components are set up taking into account some properties of the modelled hydrological process. For example, as a solution for the phase shift error in forecasting, when antecedent discharge values were the only inputs to forecast present discharge, a hybrid model was proposed by Varoonchotikul (2003). The proposed method was oriented to suppress the error of a ANN rainfall-runoff model based on the First Difference Transfer Function (FDTF: Duband et al. (1993)). The FTDF is developed from the unit hydrograph and provides an initial forecast of the future discharge which is then used as another input to the ANN.

For the shift errors other approaches, based on a time window error measure that is used to weight the objective function of a genetic algorithm optimization method, have been explored (Abrahart et al., 2007; de Vos and Rientjes, 2005, 2007).

The work by See and Openshaw (2000) mentioned above could also be interpreted as a hybrid structure. The four different approaches used for integrating conventional and data-driven based forecasting models provide a hybridized solution to the continuous river level and flood prediction problem. Although, all models were data-driven models, the best integrated solution was fuzzy system based on expert hydrological knowledge.

Additional information included in the learning process: The accuracy of DDM can be sometimes increased if additional domain knowledge (hydrological information) or specially arranged data is included in the process of model learning; however, few applications in hydrology can be found in the literature. Some examples can be mentioned:

- A multiple hybrid modelling approach was made by Hu et al. (2005), using explicit integration of hydrological prior knowledge into the learning process of a neural network. The knowledge used was the degree of wetness, which affects the runoff generated. The wetness of the catchment used the antecedent index of precipitation as an indicator. This parameter was included in the objective function, and a genetic algorithm was used for its optimization. The results show that on six out of the seven considered watersheds, the performance of the resulting model was higher than that of the overall singular neural network. The high variability in the performance of different models was demonstrated as well.

- In order to improve the extrapolation capacities of the neural networks for the modelling of rainfall-runoff it is possible to include extreme events.

This approach was made by Hettiarachchi et al. (2005), they used an estimate of an extreme flood event as input in the training process. The results show a good improvement on the extrapolation capacities.

Class 2 (D2P)

This class includes hydrological process models that either make use of the data-driven (machine learning) techniques, or have some components replaced by DDMs. The following presents some examples and sub-classes of this class of models.

DDMs as component models: Data-driven models are used to represent complex processes in physically-based modelling. An application of this can be found in the approach reported by Chen and Adams (2006), which is ideologically close to what was always the plan for this study also for predictive modelling (see Chapter 6), where data-driven models are used to replace some basin models in semi-distributed models. Their approach explores spatial modularization of conceptual hydrological models in combination with different data-driven models. The work presented by Kamp and Savenije (2007) is another example where DDM-based components are used to simplify the otherwise challenging integration of hydrological models.

Knowledge discovery by DDMs: In spite of the fact that ANN models are often treated as black boxes, exploring their internal behaviour may change this perception. Therefore some researchers try to discover the inherent hydrological knowledge by analysing the internal structures and behaviour of ANN models (Shamseldin et al., 2005; Sudheer, 2005; Sudheer et al., 2002; Wilby et al., 2003). These approaches have been undertaken by exploring relationships (correlation and regressions) between conceptual or process-based hydrological model parameters and data-driven models. Some of their results have shown that it is possible to identify hidden nodes (neurons) that in fact represent the low and high flow conditions. It is important to stress that after knowledge is identified in the data-driven model, it is possible to modify the internal structure to fit the hydrological conceptualization (Kingston et al., 2006; Sudheer and Jain, 2004). Kingston et al. (2006) presented a framework that contemplated uncertainty associated with the ANN weight vectors. However, few attempts have been made to analyse the results in terms of the physical phenomena itself (Pan and Wang, 2005).

Aside of this, clustering of data to identify physical knowledge or to find new patterns and physical concepts is a common approach to knowledge discovery (data-mining). Lauzon et al. (2006) used clustering of rain gauges created by Kohonen networks (Kohonen, 1982) to regionalize precipitation and improve ANN models.

Class 3 (DPPS)

Yet another way to look at the modelling set-up is to distinguish parallel and serial (sequential) operation of models. By "parallel" we mean that two or more models operate providing a solution for the same problem, and this architecture is typically referred to as an *ensemble*. In the serial set-up, two schemes are commonly used: (a) an "aid" model feeds information into another model (e.g. to update its states), and (b) one model corrects the output of another one (error-correcting schemes). In fact, in many operational flow forecasting systems, based on hydrological models, the serial set-up is implemented in the form of data assimilation for updating the model internal states and/or correcting its outputs.

Ensembles: These are defined as the combination of several different models responsible for the whole process under question. Commonly, in a forecast ensemble a number of models are constructed and their outputs are integrated (e.g. weighted averaged). Several authors presented and tested such an approach in hydrological forecasting. Abrahart and See (2002) performed a comprehensive study comparing six alternative methods to combine data-driven and physically-based hydrological models. Georgakakos and Krzysztofowicz (2001) analysed the advantages of multi-model ensembles where each model is a hydrological distributed model with the same structure but different parameters. The hybrid approach done by See and Openshaw (2000), used a fuzzy rule-based system and ARMA models in an ensemble using several averaging and Bayesian methods. Xiong et al. (2001) used a nonlinear combination of the forecasts of rainfall-runoff models using fuzzy logic.

For a hybrid ensemble concept different paradigms should be integrated, namely data-driven and conceptual, process-based or physically based models. A relatively recent practice explored in FLOODRELIEF project (`http://projects.dhi.dk/floodrelief`) (see, e.g., Butts et al., 2004b) is to follow the success of ensemble modelling in meteorology and other applications, and to use them in hydrological modelling as well, with the objective of reducing the output variance (uncertainty). Ensembles are researched to assess the uncertainty of individual members of the ensemble.

Some other examples of using hybrid ensemble schemes are presented below:

- Shamseldin and O'Connor (1999), introduced a combining scheme based on a linear transfer function and a weighted average method. Soil Moisture Accounting and Routing model (SMAR) and two data-driven models (linear perturbation model-LPM and Linearly Varying (LV) variable gain factor model) were integrated. The authors reported an improvement in the performance of the operational flow forecasting.

- Georgakakos et al. (2004) demonstrated that even simple ensemble models provide improvement over single modelling techniques. These authors used only simple and weighted averaging ensemble for physical based and process based models. The results show that the ensembles perform better than single models, and that both ensemble techniques have similar performance. Many researches have been making similar conclusions and reported minor differences in the performance of some ensemble techniques (Young-Oh et al., 2006).

• During the last several years the ensemble-based approach was used more and more to identify the uncertainty bounds of hydrological models (Butts et al., 2004a; Carpenter and Georgakakos, 2004; Georgakakos et al., 2004; Kim et al., 2006).

It is a common practice in complex problems to use an ensemble or a combination of models that lead to a reduction in the error variance. Although these techniques are commonly used, most of the research focuses only on one modelling paradigm, without considering if a simple integration of data-driven models or with a conceptual model may already lead to an increase in performance. This situation becomes more important when not only a simulation of a fixed time step is performed, but multiple-time-step forecasts are made. Therefore, this hybrid concept is explored by comparing a conceptual model with different combinations of models and with error corrector schemes. The comparison of different parallel and series setup is presented in Chapter 7 of this thesis.

Data assimilation: This is a model updating technique where the complementary models (DDM can be used as well) are used to identify systematic errors and/or state variations, and then correct them. Data assimilation has been extensively explored in different areas and from different perspectives. In general, data assimilation procedures have been classified by the WMO based on the variables that are modified in the feedback of the updating procedure.

• Different authors vary in quantifying the performance gains of different data assimilation techniques (Babovic and Fuhrman, 2002; Broersen and Weerts, 2005; Madsen and Skotner, 2005; Weerts and El Serafy, 2006). Madsen et al. (2000) compared global linear autoregressive models with artificial neural networks and genetic programming. Their results showed that the ANN error corrector was similar in performance to the AR models. However, the best performance is commonly obtained by the Ensemble Kalman filter and artificial neural networks, with a small difference between them. In chapter 7, a comparative analysis of the performance between the original hydrological process based model, the data-driven mode, and their comparison with a data assimilation technique is made.

2.3.2   Relationships between model classes

On the basis of the presented classification, of the approaches to hybrid model-ling, it is possible to make a diagram (Figure 2.2) with the interrelationships between some of the methods and theories related to hybrid modelling and flow forecasting.

In Figure 2.2 we see the two starting points indicating the ways of how the different modelling paradigms can be organized. In this diagram data-driven modelling combinations and hydrological modelling techniques are the basis of the hybrid modelling. From the data-driven point of view ensembles and models seem to be the logical approaches to contemplate the combinations of modelling paradigms.

The arrows, starting from square 1 and 2, show the different possible in-tegrations using single data-drive and hydrological models. In this context all the classes presented in previous section could be developed with the different modular and ensemble approaches presented.

2.4   Committee machines and modular models

The notion of a modular model (but not a hybrid one) is similar to the defi-nition of a committee machine (Haykin 1999), see Figure 2.3. The committee machine concept is broad and covers multiple machine learning approaches like modular models, mixture of experts, ensembles and others. The modular model approach has been shown to be useful in representing real-life situati-ons, and/or addresses independent and local problems (Auda and Kamel, 1998; Ronco and Gawthrop, 1995; Zhang and Govindaraju, 2000a).

CM is a term originally associated with artificial neural networks. The trai-ning and operation of committee machines is illustrated in Figure 2.3. The input data u, at time t, pass through a split unit (S, gate) which makes a selec-tion or separation of the data. A model is built for each selected or separated data stream (f_i), which will be integrated in a final module (I). This module is a unit that combines the values based on the separation or selection done in the split unit. The training process of such a model, as in any computational intelligent method, involves the feedback of the error through different models and then to their parameters.

It is possible to classify the CMs with respect to the way the splitting is performed (Solomatine, 2006):

- hard splitting followed by training multiple models, of which only one is used to predict the output for a new input vector;

- hard splitting followed by training modular models whose outputs are integrated by "soft" weighting scheme (leading, for example, to "fuzzy committees");

- statistically-driven soft splitting, used in mixtures and boosting schemes

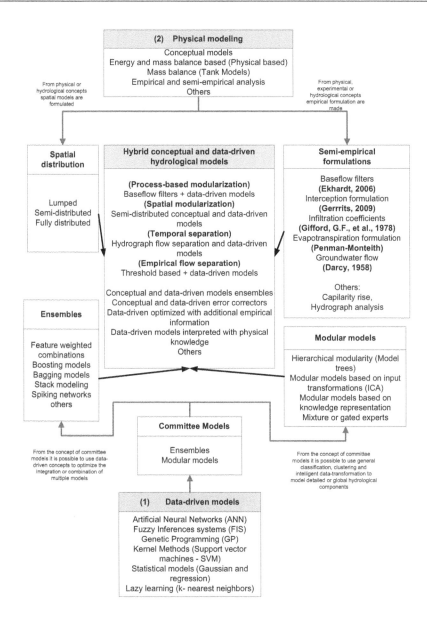

Figure 2.2: *Hybrid flow forecasting models in a data-driven modelling framework*

- no-split option leading to a collation of models; these are trained on the whole training set and their outputs are integrated using a weighting scheme where the model weights typically depend on model accuracy.

It can be seen that the approach taken in building committee machines

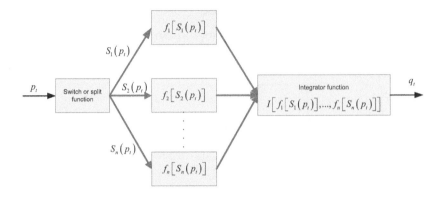

Figure 2.3: *General scheme of a committee machine model*

follows the principle of modularity, so these notions are often used interchangingly.

2.5   Measuring model performance

Models performance (typically associated with its accuracy, but sometimes also with reliability) can be measured in many different ways, and in hydrology a number of traditionally accepted measure are used. In hybrid modelling of natural processes there could be situations not handled properly by the traditional model performance measures accepted in hydrology. For example, in data-driven models it is common to find unfeasible physical results, i.e. negative values of the forecast, and this has to be handled by the performance measures. Sometimes in overall measures of error, the improvement on peaks accuracy may be hidden by high or continuous errors on low flow regions; so the use of the average performance may be misleading.

In this section several model performance measures are described; most of them are indeed widely used, like traditional statistical measures, and the use of an absolute error formulations, but also forecasting probabilistic errors are presented. The reason to use this or that error measure are discussed.

Error analysis techniques

Standard performance measures make overall assessments of the model performance. Each one has its advantages and disadvantages for performance assessment, and there is no universal agreement on what is the best measure for the calibration of a flow model. Two types of measures can be found in the different research of flow forecasting, one is based on standard statistical overall measures and the second one is the probabilistic overall error measures used in operational forecasting. Therefore a selected number of these are discussed in the following paragraph.

Statistical performance measure

A convention on the training of ANN models is the use of mean square error (MSE). To reduce the influence of the squared error information in the measure, the root of MSE is most commonly used (RMSE m^3/s, equation 2.1). This measure punishes large deviation, found often at high flows (peak flows), so when assessing the performance bias or shape may not be explicitly contemplated. This assessment could show that one model is better than other, but may have inaccurate results for low measurement values (low flows).

$$\text{RMSE} = \sqrt{\frac{\text{SSE}}{n}} \qquad (2.1)$$

$$SSE = \sum_{t=1}^{n} \left(Q_{sim,t} - Q_{obs,t}\right)^2 \qquad (2.2)$$

where $Q_{obs}(m^3/s)$ is the values of the observed discharge and $Q_{sim}(m^3/s)$ is the estimated discharge from the model. Equation 2.1 is used to answer what is the average magnitude of the forecast errors.

Sometimes it is important to compare two time series using a reference of statistical properties of measurements. Therefore, here we use root relative squared error (Witten and Frank, 2000), which compares the root square of the mean of squared errors with the standard deviation of measurement. This means that we can see if the average errors are outside of the standard deviation of measurements. This measure is sometimes expressed as percentage, so a value of 100% means that the RMSE is in the bound of the standard deviation. If the errors are much higher than these bound values the root relative squared error will be above 100%. In this sense the root relative error is a Normalized Root Mean Square, term used in this thesis (NRMSE, equation 2.3).

$$\text{NRMSE} = \frac{\sqrt{\frac{\text{SSE}}{n}}}{\sigma_{Q_{obs}}} \qquad (2.3)$$

where the $\sigma_{Q_{obs}}$ refers to the standard deviation of the observed discharge.

In hydrological sciences it is common to find what could be a variation of an error measure. The coefficient of efficiency (CoE, equation 2.4) is a normalized error measure that can be expressed as one minus n times the RMSE over standard deviation. The measure can be negative for very bad models, and the perfect model has CoE=1.

$$\text{CoE} = 1 - \frac{\sqrt{\text{SSE}}}{\sqrt{\sum_{t=1}^{n} \left(Q_{obs,t} - \bar{Q}_{obs,t}\right)^2}} \qquad (2.4)$$

$$\overline{Q}_{obs} = \frac{\sum\limits_{t=1}^{n} Q_{obs,t}}{n} \qquad (2.5)$$

Different authors point that this error measure needs to be complemented to have a measure of how much is the volume of water overestimated or underestimated. For this the relative volume error (V_E, Equation 2.6) is incorporated in the calibration of some hydrological conceptual or process-based models (e.g. R-Criterion used in the HBV model).

$$V_E = \frac{\sum\limits_{t=1}^{n} (Q_{obs,t} - Q_{sim,t})}{\sum\limits_{t=1}^{n} Q_{obs,t}} \qquad (2.6)$$

The mentioned measures can be combined. For example, the developers of the HBV model propose the following aggregated error to be minimized.

$$R_V = CoE - w\,|V_E| \qquad (2.7)$$

Where Rv is the Relative criterion based on Coefficient of Efficiency (Nash-Sutcliffe Coefficient), and w is a weighting, generally taken as 0.1 (Lindström et al., 1997).

Naïve based error analysis   The persistence index (PERS) focuses on the relationship of the model performance and the performance of the naïve (no-change) model which assumes that the forecast at each time step is equal to the current value (Kitanidis and Bras, 1980):

$$PERS = 1 - \frac{SSE}{SSE_{naive}} \qquad (2.8)$$

$$SSE_{naive} = \sum\limits_{t=1}^{n} (Q_{obs,t+L} - Q_{obs,t})^2 \qquad (2.9)$$

SSE_{naive} is a scaling factor based on the performance of the naïve model; $Q_{sim,t}$ is the DDM forecast or a process-based model simulation of the next time step, $Q_{obs,t}$ is the observed discharge at time t where $t = 1, 2, , n$; L is the lead time ($L = 1$ for one day ahead forecast); and n is the number of steps for which the model error is to be calculated. A value of PERS < 0 means that the considered model is less worthy than the naïve model (i.e. it is degrading the provided information) while 0 $<$PERS < 1 indicates that the considered model is better than the nave model (where the closer to 1 the better). Lauzon et al. (2006) suggest using PERS in cases when the discharge forecast is made on the basis of previous values.

Relative error analysis

Absolute error measures provide an equal weighted measure of the errors, so small error will have the same importance in the error formulation. The use of a relative error (RE) as an additional error measure is justified by the following. If RMSE or CoE are used, the same error value may be considered to be high in the low flow season and relatively low for the high flow season. One solution could be to weigh the error values differently for different seasons, but such an approach will still depend on the objective identification of the low and high flow regions. Another solution is to use RE, which automatically takes into account the value of the measured variable, so that a value of RE corresponding to large absolute errors in the case of low flows is large while it will be relatively lower in the case of high flows (Figure 2.4).

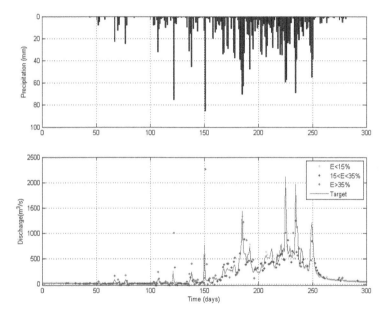

Figure 2.4: *Relative errors found in the rainfall-runoff ANN model of the Bagmati catchment in Nepal*

In this study an overall *RE* (Equation 2.10) is used to identify the percentage of samples belonging to one of three groups: "low relative error" with *RE* less than 15%, "medium error" with *RE* between 15 and 35%, and "high error" with *RE* higher than 35%. The ranges were determined after experiments with the two trial models. The low error value is expected to cover possible measurement errors that could be around 20% (Beven, 2003).

$$\text{RE}_t = \frac{|Q_{obs,t} - Q_{sim,t}|}{Q_{obs,t}} 100\% \tag{2.10}$$

Forecast performance measures

Most of the previously mentioned statistical performance measures reveal information about the following characteristics of the generated forecast:

- Bias - the correspondence between the mean forecast and mean observation.

- Association - the strength of the linear relationship between the forecasts and observations (often analysed by the correlation analysis of the outputs)

- Accuracy - the level of agreement between the forecast and the truth (as represented by observations). The difference between the forecast and the observation is the error. The lower the errors, the greater the accuracy.

However, in operational forecasting it is more common to look at critical results that are defined by a threshold. In this sense in meteorological science it is common to use a measure based on what is called a skill. The skill is a relative measure of the accuracy of the forecast over some reference forecast. This error measures can be seen as a probability of hitting or missing certain target or threshold (Table 2.2). For this a contingency table is built on the basis of the number of occurrences and non-occurrences.

Table 2.2: *Contingency table*

		Observed		Total
		Yes	No	Total
Forecast	*Yes*	Hits (Hi)	False alarms (Fa)	Forecast *yes*
	No	Misses (Mi)	Correct negatives (CoN)	Forecast *no*
Total		Observed *yes*	Observed *no*	Total *To*

The total numbers of observed and forecast are classified as hits or missed values, i.e. occurrence of an event, or non-occurrence. Forecast is perfect if hits are found and there are no misses or false alarms.

To analyse the overall results it is important to know what fraction of the forecasts is correct. This *accuracy* value will vary from 0 to 1, and can be calculated as shown in Equation 2.11. This measure has a drawback, since it is heavily influenced by the most common category, usually "no event" in the case of rare precipitation events or floods.

$$Accuracy = \frac{Hi + CoN}{To} \tag{2.11}$$

The Bias can be calculated by determining the ratio of forecast *yes* (false alarm and hits), compared to the observed *yes* (hits+misses), Equation 2.12. This rate value varies from 0 to infinity, with a perfect forecast in 1. If BIAS<1 we have underforecast, and if BIAS>1 overforecast events. The bias does not measure how well the forecast corresponds to the observations, if only measures relative frequencies.

$$Bias = \frac{Hi + Fa}{Hi + Mi} \tag{2.12}$$

The most common measure is the probability of detecting (POD) an event (Exceeding of a threshold, Table 2.2), which in flow forecasting is a flood, Equation 2.13. For this the table should answer "What fraction of the observed "yes" events were correctly forecast?" This measure has perfect forecast when with a value of 1. However, as it can be seen, it is only sensitive to hits, and ignores false alarms.

$$POD = \frac{Hi}{Hi + Mi} \tag{2.13}$$

The probability of detection needs to be combined with it opposite situation of failure (Equation 2.14. The false alarm ratio "What fraction of the predicted "yes" events actually did not occur?". As the POD, the FAR goes from 0 to 1 and it is complementary in the sense that it ignores the misses.

$$FAR = \frac{Fa}{Hi + Fa} \tag{2.14}$$

2.6 Discussion and conclusions

Recently a number of modelling approaches, including hydrological conceptual and data-driven models, have been explored. It raised the need to have better methods for their integration. The importance of the awareness of modelling approaches and available tools is becoming an issue for new researchers in this area. This chapter introduced a classification of hybrid models, with the objective to group and generalize methods for integration. The description of each class is done by referring to example modelling approaches that fit into the classification scheme presented.

The reviewed literature presents a number of successful examples of using various modular and hybrid approaches. Although the reported efforts already show the advantages of these techniques in hydrological forecasting, we hardly could find attempts to classify these approaches, and consider them in a more-or-less general framework. Most of these techniques seem to be appearing through interpretation of advantages (higher accuracy) over a benchmark model, not from a perspective of comparing modelling paradigms.

A general framework of different modelling alternatives is presented to complement the developed classification of hybrid models. The framework is the basis for different ways of integrating hydrological conceptual or process-based models and data-driven models. It could be said that this framework can be the starting point for assessing relevant and non-relevant modelling approaches. Having an overall concept of different modelling approaches and their possible interactions may help on the formulation of new optimal hybrid models. However, the framework presented is not complete; it will certainly evolve and grow.

The notions and methods related to committee machine and modular models have been introduced and will serve as the basis for hybrid modelling in this thesis. The novelty of the approaches presented in the following chapters is the use of simple hydrological empirical data separation techniques that challenges single-model approach, and automatic clustering algorithms. Other ways to integrate models integrations will be evaluated and their advantages in terms of knowledge representation will be discussed.

OPTIMAL MODULARIZATION OF DATA-DRIVEN MODELS

In the previous chapter, the modularization of data-driven models in the context of hybrid modelling was introduced. Different modular models were reviewed and it was observed that often there is insufficient use of hydrological knowledge in the modularization process. Therefore, in this chapter we explore alternative methods for building modular models that are based on the explicit use of hydrological knowledge (in particular, separate consideration of base- and excess flow), and also use data partitioning algorithms. The developed modelling techniques will be applied in Chapter 5, where they will be compared with a single overall data-driven model, and with conceptual models. Application of the principle of modularization to separate modelling of processes that relate to different areas in space is considered here as well; it is further applied in Chapter 6. There is also a possibility of multi-objective optimization of a model involving various modularization schemes, and this will be mentioned as well (however not explored). This chapter is to a large extent based on the methodology published by Corzo and Solomatine (2007a,b) and Corzo et al. (2009a).

3.1 Introduction

Traditionally, modellers were, and often still are, trying to build a general, all-encompassing model of a studied natural phenomenon. Hydrological forecasting models that involve the use of data-driven techniques are not exceptions in this sense: they tend to be developed on the basis of using a comprehensive (global) model that covers all the processes in a basin (Abrahart and See, 2000, 2002; ASCE, 2000a,b; Dawson et al., 2002; Dibike and Abbott, 1999; Dibike and Solomatine, 2001). Such models (very often these are artificial neural networks, ANN) do not encapsulate the knowledge that experts may have about the studied system, and in some cases suffer from low accuracy in extrapolation. In many applications of data-driven models, the hydrological knowledge is " embedded " in the model via the proper analysis of the input/output structure and choice of adequate input variables (Bowden et al., 2005a). However, much more can be done to incorporate domain knowledge into these models, as mentioned in Chapter 2.

One of the ways of doing this is to try to discover different physically interpretable regimes of a modelled process (or sub-processes), and to build separate specialized ("local") models for each of them, either process (physically-based) models, or data-driven ones. Such an approach is seen as one of the ways of including hydrological knowledge and improving the model's performance. In order to combine the local models one may refer to the methods developed in machine learning (Haykin, 1999; Kuncheva, 2004), and, possibly, to combine them, like it is done in the so-called fuzzy committee approach (Solomatine, 2006; Solomatine and Corzo, 2006). In this latter paper we also presented one possible classification of modular models.

Lately, a number of studies were reported where such an approach was undertaken (often being named differently, however). Solomatine and Xue (2004) applied an approach where separate ANN and M5 model-tree basin models were built for various hydrological regimes (identified on the basis of hydrological domain knowledge). Anctil and Tapé (2004) applied wavelet and Fourier transforms to the identification of high and low flows based on their frequency patterns. Some attempts have been made to find correlations between ANN components and processes in a conceptual model (Wilby et al., 2003). Solomatine and Siek (2006) presented the M5flex algorithm where a domain expert (e.g., a hydrologist) is given more freedom in influencing the process of building a machine learning model. Mitchell (1998) used a combination of ANNs for forecasting daily streamflow: different networks were trained on the data subsets determined by applying either a threshold discharge value, or clustering in the space of inputs (several lagged discharges, but no rainfall data, however). Jain and Srinivasulu (2006); Wang (2006) also applied decomposition of the flow hydrograph by a certain threshold value and then built separate ANNs for low and high flow regimes.

All the previous mentioned studies demonstrated the higher accuracy of the modular models approach when compared to the models built to represent all possible regimes of the modelled system (such models are referred to herein as *global models*). However, in terms of contemplating the physical principles, more can be done.

If we want to follow the idea of a modular approach, there is the possibility to take a somewhat deeper view of the underlying sub-processes to be modelled for accurate flow forecasting. In river basin modelling, a typical approach would be to identify baseflow and the direct runoff (also called excess runoff or excess flow). Such an approach was described by (Solomatine and Corzo, 2006), the present study develops it further.

The main idea used here is to use specialized algorithms for the hydrograph analysis to separate baseflow from excess flow, form training data sets and build local ANN-based models for each component. The focus is on optimization of the model structures, and of the parameters of the data separation and combination algorithms.

3.2 Methodology of modular modelling

The problem of hydrological modelling of a basin considered in this thesis is characterized by precipitation and discharge measured at different moments in time in the past (which can be seen as multivariate time series), and the response of the river basin represented by the forecast of the discharge (flow) hours or days ahead. This can be expressed as follows:

$$Q_{t+T} = f(R_t, R_{t-1}, R_{t-2}...R_{t-L}, Q_t.....Q_{t-M}) \tag{3.1}$$

where Q_{t+T} is the discharge in m^3/s, R is rainfall in mm and t the actual time where the forecast is made, the lags L for precipitation and M for discharge are obtained through model optimization (these can be different for various forecast horizons T); f is the data-driven regression model, and T is the forecast horizon (e.g. 1 day according to the available data).

A general structure of a modular model is shown in Figure 3.1, where each specialized (local) model would represent one sub-process or regime. In computational intelligence, the overall model is often called a committee machine (Haykin, 1999).

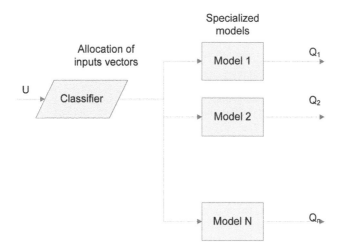

Figure 3.1: *A modular model based on local specialized models*

A modular model, as any other DDM, undergoes the normal training, validation and testing (operation) stages. However, due to the existence of several model components, a modular model also has an additional element – a classifier (Figure 3.1) that during operation (testing) would attribute data to one or more models. Two phases of the process can be distinguished. First, the classifier should be built (Figure 3.2) and trained, and second, it will be used to allocate input vectors to models during operation (testing) phase. In different versions of modular models either all models are run (forming thus an ensemble, or a committee), or only one of them. Depending on this, outputs of the models are either combined (using some averaging or weighting scheme), or only the output of the single activated model is used as the model output.

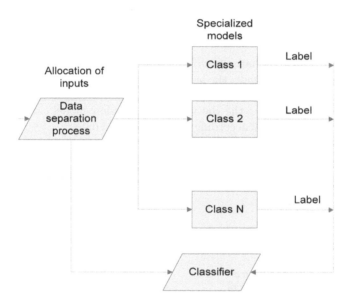

Figure 3.2: *A modular model based on local specialized models*

As mentioned before, the main idea used for modularization considered in this chapter is based on baseflow separation (Figure 3.3): instead of building one model responsible for representing the water flow for all regimes, two models are built. One model simulates the baseflow, and the other one simulates the direct runoff or the total flow. These two models work sequentially or in parallel.

A conventional engineering approach to find the baseflow is to generate and hydrograph and make a geometrical division as presented in Figure 3.3. It can be said that before time moment t_s, the hydrograph consists only of baseflow; between t_s and t_r the baseflow and direct runoff together form the the total

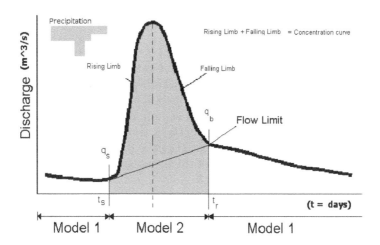

Figure 3.3: *Flow separation used to build local specialized models (t_s = beginning of storm, t_r = beginning of recession).*

flow; after that the flow reaches a baseflow condition again. Although this is a quite old formulation, and it is well known that multiple flow components from different soil layers and other phenomena compose this curve, the principle is still universally valid. This can be interpreted in a more general way and assumed that the water system has two states: the low flow state (baseflow), and the high flow state.

Even though the baseflow formulation presented for the hydrograph analysis is simple, the problem here is to find the moments when the flow regime changes from low flow (baseflow only) to the regime where high flow (direct runoff + baseflow) is present as well. The graphical formulation need to be encoded into mathematical representations of the behaviour of the hydrograph. Also, in a more general sense the detection of changes in state of a highly variable system is still an ongoing issue in many areas (e.g. Valdés and Bonhjam-Carter, 2005).

There are various ways to partition the data into blocks to be used to train different models. In this chapter, when considering the separate modelling of low and high flow we will be using three ways of doing partitioning: 1) based on clustering, 2) based on identification of sub-processes as a result of hydro-logical analysis (referred to as "process-based"), and 3) based on separation of the processes in time using a filtering algorithm (which, strictly speaking may be seen as a simplified version of the process-based partitioning). We follow the definitions of baseflow given by Hall and Anderson (2002) for time-based separation, and by Chapman (2003, 1999) for process-based separation. Note that there are also other interpretations of how to define the baseflow (Beven, 2003; Uhlenbrook et al., 2002).

3.3   Modularization using clustering (MM1)

This modular scheme is based on the application of a clustering algorithm to the vectors (instances) in the space of input variables. The modular models are trained as shown in Figure 3.4. The clustering process returns the index of the vectors (I_x) for each class (C_x). We used a k-means algorithm to identify two clusters, and conducted a number of experiments to identify best distance metric for the clustering scheme. In this thesis three different distance metrics are tested: cityblock (the sum of the absolute differences), Euclidean, and cosine (distance is one minus cosine of the included angle between points).

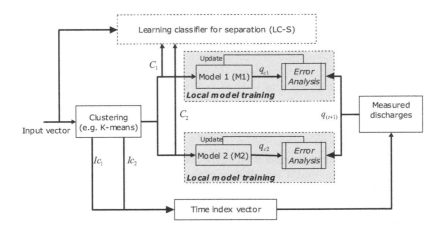

Figure 3.4: *Modular model with a classifier trained on the data from clustering by k-means method*

Training a classifier for data separation (LC-S):   To use the resulting model, a learning algorithm has to be applied to build a classifier that would classify the new examples according to the clusters identified by the k-means algorithm. For this purpose the following classification algorithms were explored:

- Fisher's linear discriminant. It is a classical classification method (Duda, 1996) that projects high-dimensional data onto a line and performs classification in this one-dimensional space. The projection maximizes the distance between the means of the two classes while minimizing the variance within each class. This defines the Fisher criterion, which is maximized over all linear projections. Maximizing this criterion yields a closed form solution that involves the inverse of a covariance-like matrix.

- Decision trees (DT) or regression trees (RT). The important feature of these classifiers is that they can be easily interpreted by human experts since they are actually represented as sets of if-then-else rules. The algorithm to build regression trees proposed by Breiman et al. (1984) is a regression algorithm in which the input space is progressively partitioned into subsets by hyperplanes $x_i = A$ (where x_i is one of the model inputs, and i and A are chosen by exhaustive search and criteria based on maximisation of information at each split).

A leaf in such a regression tree is associated with an average output value of the instances sorted down to it (zero-order model). In the considered classification problem there are only two classes (labelled as "baseflow" and "non-baseflow") so data for training RT had to undergo transformation, and during the operation its output has to be converted to a class label. RT appeared to be an accurate classifier.

- Probabilistic Neural Networks (PNN). The approach taken in this method follows in general Bayesian theory, and uses Parzen Estimators (Parzen, 1962) to construct the prior probability density functions. Details can be found in Specht (1990) and Duda (1996). Data to train PNN has to be transformed in a way similar to the one described for RT above.

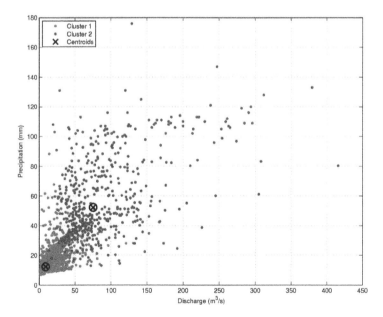

Figure 3.5: *K-means clustering results for the precipitation and discharge time series from 1 Jan 1989 to 28 dec 1995 (Ourthe river basin, Belgium)*

Under certain assumptions or conditions, the use of a clustering technique can be interpreted as an automatic identification of the ongoing regimes (Geva, 1999). There is no guarantee, of course, that a direct relationship between clusters and identifiable hydrological regimes would be discovered (Figure 3.6). The clustering methodology is just taken as a typical example of automatic separation methods for modular models. Clustering-related experiments should be seen as a demonstration of a possibility of such an approach.

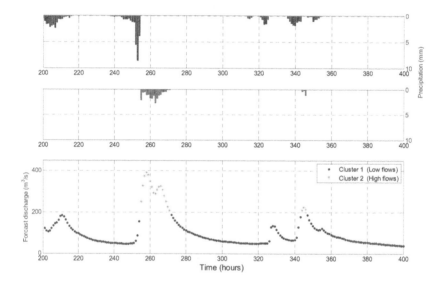

Figure 3.6: *Fragment of the hourly hydrograph with a cluster-based separation (Sieve case study)*

Standard $k-$means algorithm (Spath, 1985) was used to find groups of input vectors of discharge and precipitation. This algorithm is based on a two-phase iterative process, which minimizes the sum of point-to-centroid distances, summed over a number of clusters (k). At each iteration, points are assigned to their nearest cluster centre (chosen randomly at the very first iteration), followed by recalculation of cluster centres. The number of clusters has to be chosen *a priori*.

There is a wide variety of distance metrics to be used and for the purpose of this study the city-block (Manhattan) was selected. The errors of the models built using several distance metrics were compared and Manhattan metric appeared to result in the lowest error. The distance metric used to build the cluster can be described as follows. Given an (mn) data matrix X, the various distances between the vector x_r ($r = row$) and x_s ($s = col$) are:

$$d_{rs} = \sum_{j=1}^{n} |x_{rj} - x_{sj}| \qquad (3.2)$$

Note that a separate classifier for data splitting has to be built to serve as a splitting unit during operation (Equation 3.2); its purpose is to attribute new examples to an appropriate model. The training data for such a classifier is constructed in the following way: the input data are the same as the input data for models ANN model (M1) and (M2), and the output data are the class labels corresponding to the identified clusters (Figure 3.2 and Figure 3.4).

The result of applying the described procedure is a model termed as MM1 consisting of:

a. ANN model 1 and ANN model 2 representing low flows and total flow respectively, and trained separately on the data subsets corresponding to the identified clusters;

b. Classification model for data separation during operation or testing (referred on Figure 3.4 as LC-S).

3.4   Modularization using sub-process identification (MM2)

Instead of grouping the input vectors, one may explicitly use hydrological domain knowledge to partition the data into groups that would be modelled separately. The flow of water through the basin is heterogeneous, follows various routes and, in hydrological analysis, and it is often right to talk about the baseflow and the direct runoff (Hall, 1968; McCuen, 1998). Classical hydrograph baseflow separation analysis is in fact a graphical semi-empirical technique that splits the discharge values based on the measurement of discharge and precipitation (Fig. 3.3). In it, the values q_s and q_b of flow during the multiple storms are found, and the starting (t_s) and ending (t_b) of a storm phenomenon are identified.

The traditional "constant slope" method (McCuen, 1998) is manual: the beginning of the storm is identified as the point where the discharge is minimum, and the end of the storm corresponds to an inflection point. These two points are connected thus determining the sought baseflow area, and the slope of this line is recorded. Recently a number of other, simpler, methods have been introduced (Engel and Kyoung, 2005; Sloto and Cruise, 1996). However, in the MM2 scheme the "constant slope" method is used as the main foundation for building the separation algorithm. However, a modification that allowed for an easier algorithmic implementation was used: instead of looking for a hydrograph minimum, it is based only on the identified inflection points. To connect to the point where the direct flow finishes, a linear interpolation can be drawn from one inflection point to an inflection point at the end of the storm period (instead of connecting the minimum to an inflection point). The resulting line was found to be almost the same as the one identified by the traditional manual method (Figure 3.7).

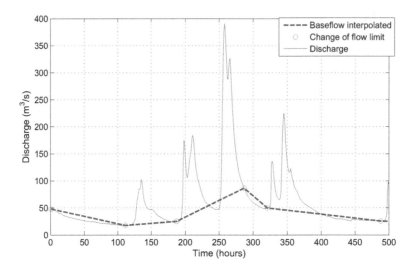

Figure 3.7: *Baseflow separation results with the inflection points implementation*

Algorithmic implementation The accurate determination of the inflection point by analysing the time series derivative estimates requires analysis of the complete event. Due to the high variability of data in a typical hydrological time series, the implementation of this method is not straightforward since there are too many inflection points even in short time intervals. One way to resolve this is smoothing of the time series and then identifying the inflection points by analysing the second derivatives. However, in our experiments such analysis still result in finding many inflection points, and it is necessary to focus only on those that correspond to the beginning and ending of an event.

To remove such "false" points, it was chosen to apply the method presented by Sloto and Cruise (1996) (who, however, applied it to the minimum points rather than to inflection ones). Their idea was to remove points that lie on the hydrograph within the period of a storm event; however, to identify this period is of course not a trivial problem. Fortunately, there are empirical methods known to do this, and Sloto and Cruise used the method of Linsley et al. (1982). The latter suggested that the average storm duration should be close to two times the number N of days between the peak flow and the end of the direct runoff. To assess N the following equation is used: $N = A^p$, where A is the basin area, and p varies depending on the basin characteristics. It was used $p = 0.2$ based on the recommendations found in (McCuen, 1998). This approach is not very accurate, but appears to work well for the purpose of the baseflow separation algorithm.

The time-based baseflow separation algorithm was implemented as follows:

1. **Smoothen the data**: This step is made to ease the identification of inflection points. The moving average filter is used; the span n of the filter can be changed according to the hydrological conditions of the case study (i.e. concentration time):

$$Q_{s_t} = \frac{1}{(2n + 1)} \left(Q_{t+n} + Q_{t+n-1} + \cdots + Q_{t-n} \right) \tag{3.3}$$

where Q_t is original discharge; Q_{s_t} is the smoothed discharge; and n is the filter span.

2. **Find the inflection points**: The inflection point is defined as the point where the second derivative of the discharge is zero, as follows:

$$\frac{\partial^2 Q_s}{\partial t^2} \approx \frac{\Delta^2 Q_s}{\Delta t^2} = \frac{\Delta}{\Delta t} \left(\frac{\Delta Q_s}{\Delta t} \right) = 0 \tag{3.4}$$

3. **Remove the "false" inflection points** using the notion of the average storm duration, as described above.

4. **Separation.** Finally, a (virtual) line is drawn between the inflection points, which separates the baseflow from direct runoff and thus separates the two regimes: one when only the baseflow is present and the other one when the baseflow is accompanied by the direct runoff (thus constituting the total flow). This line also graphically represents the switching rule for the two predictive models. Algorithmically, a linear separation model is used (Figure 3.8).

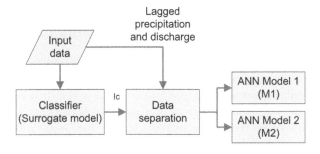

Figure 3.8: *MM2 modular scheme in operation*

The presented baseflow separation method needs the data corresponding to the whole storm, and such data is available during training, so the algorithm can be used to separate the data and train two separate models which are referred to as M1 (ANN model 1) and M2 (ANN model 2).

5. **Generate data for the LB-S model.** Apply steps 1-3 to the historical data and generate enough data for training (Figure 3.8).

6. **Train LB-S model.** Use the generated data to train the a surrogate model (a machine learning model, e.g. ANN) that would predict the location of inflection points in the operation phase of MM2.

Learning baseflow separation (LB-S) using a surrogate model. The method of base flow separation requires the data set corresponding to the whole storm event. Such data is available during training, so the two models can be trained as described. However, during operation, the algorithm cannot be applied since it is based on processing of the data characterizing the whole storm event. Hence, a surrogate model replicating the baseflow separation algorithm (Figure 3.8) is needed. The learning process follows the same principles presented in the clustering modular model scheme (Figure 3.1 and 3.2), but in this case the data classified are obtained from the presented semi-empirical baseflow separation method.

The experimental results presented later in chapter 5 are based on the use of the linear discriminate classifier. During validation and operation phase the trained classifier LB-S identifies the inflection points used for attributing the new examples to one of the two trained ANN models.

The result of applying the described procedure is the MM2 model consisting of:

a. ANN model 1 (M1) and ANN model 2 (M2) trained separately to model baseflow and direct flow;

b. Model LB-S that in the operation (or testing) phase identifies the inflection points used for data splitting.

3.5 Modularization using time-based partitioning (MM3)

The traditional baseflow separation methods mentioned above cannot be effectively used when separations are to be undertaken on a long continuous record of streamflows, rather than just a few storm period hydrographs. This has led to the development of a class of algorithms sometimes referred to as "numerical". Relatively recent research has applied flow separation filters that consider one or two variables in recursive algorithms (Arnold and Allen, 1999; Chapman, 2003). However, these authors define baseflow slightly differently, assuming that even in the periods of low flow there are two components of flow which can be interpreted as direct runoff and baseflow. The method used in this study is based on the baseflow recursive filter (Ekhardt, 2005). Ekhardt (2005) compared many of the existing baseflow filtering algorithms and proposed the following equation:

$$q_{b_t} = \frac{(1 - \mathrm{BFI}_{\max})\,a q_{b(t-1)} + (1 - a)\,\mathrm{BFI}_{\max} Q_t}{1 - a\mathrm{BFI}_{\max}} \tag{3.5}$$

where q_{b_t} is the baseflow at time step t; $q_{b_{t-1}}$ is the baseflow at the previous time step; Q_t is the measured total flow; BFI_{max} is a constant that can be interpreted as the maximum baseflow index; and a is a filtering coefficient, or recession constant. The three coefficients q_{b_0}, BFI_{max} and a are unknown, and there is no commonly accepted method to identify them. In principle, identification of coefficients is based on trial and error and sometimes it is possible to use the recession curve coefficients. In this thesis the three mentioned coefficients are found through an optimization process. The goal of optimization is to find such coefficients that ensure the overall modular model has the best performance (minimum error).

Modelling scheme based on time-based partitioning   (Figure 3.9) The ultimate goal is to predict the total flow Q_{t+1} at the next time step. The baseflow filter (Equation 3.5) separates the flow into two components and they are fed as inputs to the models. Both models have to be trained on the basis of the measured data, and the model structure has to be optimized as well. The optimization of MLP ANN had the objective to find as well the optimal number of hidden nodes. During testing and operation phase the optimized baseflow filter (Equation 3.5) is used to attribute the new examples to one of the two trained ANN models.

Filter optimization   The filter parameters q_{bf_0}, BFI_{max} and a are to be optimized. Note that during the optimization, in order to calculate the error (RMSE) for every new parameter vector, two ANNs are to be trained, so the process can be computationally expensive. To be able to evaluate a fair "trade off" in the performance of the local specialized models and the overall model error, a weighted objective function is used; it involves a weighted function with a high weight for the overall model error and a low weight for the baseflow component model error (errors were measured on the training subsets):

$$E_T = w_0 \mathrm{RMSE}_{(Overall)} + w_1 \mathrm{RMSE}_{M1} + w_2 \mathrm{RMSE}_{M2} \tag{3.6}$$

where $M1$ and $M2$ refer to the specialized models of direct flow and baseflow, respectively; and wi are the corresponding weights. The weights should be subject to optimization as well. In the presented study they were however selected on the basis of a (limited) number of experiments with the subsequent visual inspection of the resulting hydrograph, and following the assumption that our main objective is to forecast the flood situation. The resulting values

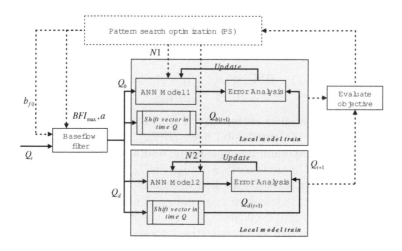

Figure 3.9: *Optimization of the process-based separation model (MM3)*

were 0.6, 0.3 and 0.1 respectively. The objective function (related to the model error to be minimized) is calculated by the following procedure:

1. generate a random vector b_0, BFI_{max}, a, number of hidden nodes in each ANN;

2. run equation (3.5) to perform the baseflow separation, generating two different training sets;

3. train two different ANN models $M1$ and $M2$ of direct and baseflow respectively;

4. calculate the overall error E_T using equation (3.6) (total flow is found as the sum of the models outputs).

For global optimization various methods can be used, and since the objective function is not known analytically the so-called direct optimization methods should be used. This is a class of methods that do not use gradients explicitly; for example all randomized search algorithms (like GA) can be attributed to direct search methods; they are used in cases when the objective function is not differentiable, stochastic, or discontinuous. Certain direct search algorithms use some sort of gradient assessment (using finite-difference approximation) search for a set of points around the current point, looking for one with the lower value of objective function compared to the current point. Some trials and errors were explored using direct search methods such as randomized search, GA, adaptive cluster covering (ACCO, Solomatine (1999)), and Generalized Pattern Search (GPS). In the present chapter GA and GPS are applied based

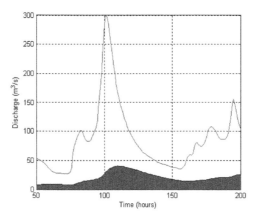

Figure 3.10: *The measured total flow (solid line) and the baseflow filter output (filled area) (Sieve case study)*

on the implementation provided in MATLAB (e.g. Genetic Algorithm and Direct Search Toolbox).

In comparison with a popular GA, GPS is probably less known, so a brief description of GPS follows. At each step, the algorithm searches for a set of points, called a mesh, around the current point the point computed at the previous step of the algorithm. The mesh is formed by adding the current point to a scalar multiple of a set of vectors called a pattern. If the pattern search algorithm finds a point in the mesh that improves the objective function at the current point, the new point becomes the current point at the next step of the algorithm. The GPS algorithm uses fixed direction vector to define the mesh. The size of the mesh changes at each iteration. The update of the mesh is based on the difference between the current point and the new point. For further details the reader is referred to Xiaokun et al. (1991) and Abramson et al. (2004).

3.6   Modularization using spatial-based partitioning

The principle of modularization can be also used to separate modelling of processes that relate to different geographical locations or areas in space. Spatial modelling have become quite common in the last decades, however, the use of multiple models in a spatial ensemble is still not a common practice. This approach will be considered not in relation to a data-driven model, but to a process-based one, or a hybrid model. However, it is worth mentioning it in this chapter (mainly devoted to data-driven models) for completeness.

Consider, for example, a hydrological model of a river basin consisting of

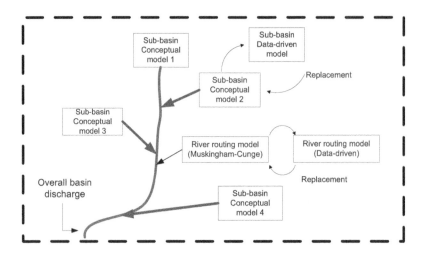

Figure 3.11: *Spatial models in operational forecasting environment*

a number of local models of sub-basins. It may happen that some of these sub-models need replacement since they are not accurate or because of some other reason. In this study the possibilities of replacing some of the least accurate of the (geographically) local models (modules) in a (semi-)distributed process-based model are explored. Focus is on replacing the lumped conceptual (sub-)models by the data-driven ones. The approach is based on selective replacements that follow the main river path. The different sub-basin models from upstream to downstream have different influence on the overall model accuracy for the extreme and normal flow situations downstream, and by studying such influence the decisions about the components considered for improvement or replacement by other types of models can be made.

Selection of the model components to be replaced can be also based on expert judgement or some simple rules (Figure 3.11). These rules can be based on the known issues related to models used in a forecasting system, or the main objective of forecasting in a particular case. For example, if the phenomena like fast floods (which are driven by precipitation) are to be modelled, the process-based models could be the first choice. On the other hand, floods driven by slow increase in the discharge may be better modelled by data-driven models, or models where some conceptual or even process-based components (models of sub-basins) are replaced by data-driven models (such models may be referred as hybrid). The details of the modularization using spatial-based partitioning are presented in chapter 6; on the Meuse river basin.

3.7   Optimal combination of modularization schemes

The different modularization schemes can be combined, to form an overall model, and there are various ways of doing this. Distributed and semi-distributed models could have different structures and hence allow for multiple ways of improving them through various types of modularization. Different flow regimes can be identified separately per sub-basin, and then all models combined. It would be reasonable to pose this process as an optimization problem, where various desired aspects of the overall modelling effort would be taken into account. Such aspects can be formulated as objective functions to ensure, for example:

- Obtain the best overall performance of the model in forecasting downstream discharge;

- Obtain the best performance of sub-models for particular regions of a watershed;

- Obtain the best representation of particular types of flow, etc.

Some of the objective functions could be also re-formulated as constraints. An appropriate setting for this would be a multi-objective optimization scheme allowing for the detailed analysis of trade-offs.

3.8   Conclusions

In this chapter the main principles of building modular models based on the explicit use of hydrological knowledge and using data partitioning algorithms have been presented and explored. This was done by (a) employing clustering algorithms, and (b) exploring the possibility of separating base- and excess flow using two different approaches. On the basis of such data partitioning separate models for each sub-process have been built. New algorithms and modelling have been developed and tested. Applications of the modularization principle to separate modelling of processes that relate to different areas in space have been considered here as well. The possibilities for overall optimization of various modularization schemes combined have been highlighted as well.

This chapter provides the principles and algorithms to be employed in Chapters 5 and 6 for particular case studies.

BUILDING DATA-DRIVEN HYDROLOGICAL MODELS: DATA ISSUES

In Chapter 2 the conceptual framework for hybrid modelling has been presented. Data-driven modelling (DDM) constitutes an important part of this framework, and deserves more detailed presentation. The main objective of this chapter is to analyze the data-driven modelling process and the most important factors typically considered when building such models. The machine learning (computational intelligence) techniques used (artificial neural networks, Model Trees and Support Vector Machines) are characterized here, but (briefly) explained in Appendix B. Part of results presented here has been worked out and discussed in Elshorbagy et al. (2009a,b).

4.1 Introduction

Data-driven modelling is becoming a increasingly popular as a complementary technology for modelling complex natural phenomena (Dawson et al., 2005; Solomatine and Price, 2004). Most of these techniques use the advances in machine learning (ML) and computational intelligence (CI);practically all known techniques have been tried in hydrological research. They have been researched and to some extent accepted as an accurate alternative to physical based models (ASCE (2000b); Solomatine (2005); Solomatine and Dulal (2003)). The general form of the most widely used structure of a data-driven rainfall-runoff model is presented in the previous chapter by Eq. 3.1.

It is important to provide some guidance to the practitioners that are supposed to try the new models designed by researchers. It is impossible to have the same results from every model used to model a particular natural phenomenon, so some rational choice procedures based on the relevance of the measured variables, model performance and other factors would be needed in the process of selecting the model to be really used. In this chapter special attention is given to the choice of the input variables, data partitioning issues and the corresponding variability in models performance. It is suggested that the presented procedures should become part of the overall modelling framework used in operational forecasting.

To illustrate these, the Ourthe river basin (Belgium) is used as the case study. The reason to use this example is that it is a part of the Meuse river basin, which is major case study in the subsequent chapters.

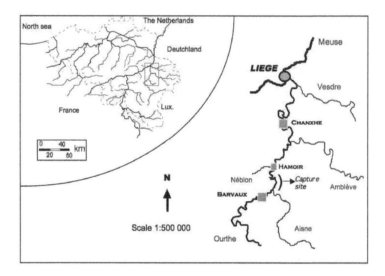

Figure 4.1: *Localization of the Ourthe river basin*

4.2   Case study (Ourthe river basin - Belgium)

The hydrological characteristics of the Ourthe river basin have been explored widely in different publications (de Wit et al., 2007a; Tu et al., 2005). The description of the whole Meuse basin will be presented in Chapter 6.

The Ourthe is the most important Meuse tributary in terms of flood forecasting. In its upper course the Ourthe consists of two branches: the Ourthe Occidental and the Ourthe Oriental, uniting near Nisramont. After the confluence of the two there is a dam, and after the Ourthe flows roughly in north-west direction. The most important tributaries of the river Ourthe are the Amblève and the Vesdre (Figure 4.1). Near Comblain-au-Pont the Amblève joins the Ourthe and near Angleur the Ourthe also receives the Vesdre. Counting from the source of the Ourthe Occidental the Ourthe is approximately 175 km long. It is located in the Ardennes Mountains in the Walloon region (Belgium).

Of all the Meuse sub-basins, the Ourthe has the largest area ($3,626\ km^2$ at Angleur). It is a typically Ardennes river, in a mountainous region, and therefore has great discharges rising fast. From a hydrologic point of view the tributaries Ambleve and Vesdre are so impotant that in Chapter 6 they are considered and modelled separately. The main reason for this can be attributed to: (a) the discharges can be greater than that of many other Meuse tributaries, and (b) the mouths of both tributaries are to be found near the mouth of the Ourthe.

A complete and detailed information of this based in the context of flood forecasting for the Meuse river basin can be found at de Wit et al. (2007a) and Berger (1992).

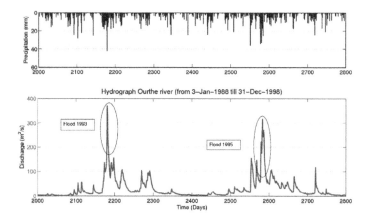

Figure 4.2: *Hydrograph of the Ourthe river basin, from 3-Jan-1988 till 31-Dec-1998*

Precipitation, temperature and discharge information was used in the analysis of this chapter. Data from 3-Jan-1988 till 31-Dec-1998 was used (4016 samples). Precipitation and temperature data were obtained from catchment averages. These data is plot in Figure 4.2; the flood situations in the Meuse around 1993 and 1995 have been highlighted with a circle.

4.3 Procedure of data-driven modelling

Data-driven (machine learning) models are typically built following a generally accepted procedure (Mitchell, 1998; Pyle, 1999; Solomatine, 2005):

- Explore the problem and solution spaces (e.g. state the problem)

- What is the expected result, and how the result will be used?

- Select the input and output variables

- Specify the appropriate modelling methods and choose the tools (e.g. software and algorithms)

- Prepare and analyze the data

- Build (e.g. calibrate or train) the model

- Test the model

- Apply the model and evaluate the results.

In reality the process of modelling is not linear, but continuous with feedback loops. For example, the lack of particular data may lead to a change of a modelling method selected. For these processes there are a set of "golden rules", here mentioned as a sort of a check list helping a modeller in the process of model building based on Solomatine (2008):

1. Clearly define the problem that the model will help to solve.

2. Specify the expected solution for the problem.

3. Define how the solution delivered is going to be used in practice.

4. Learn the problem, collect the physics (domain knowledge) and understand them.

5. Let the problem drive the modelling, including the tool selection, data preparation, etc.

6. Take the best tool for the job, not just a job you can do with the available tool.

7. Define clearly assumptions (do not just assume, but discuss them with the domain knowledge experts).

8. Refine the model iteratively (try different things until the model seems as good as it is going to get).

9. Make the model as simple as possible, but no simpler. In machine learning (ML) it is referred to as the Minimum Description Length principle saying that the best model is one that is the smallest length (including the information to specify both the form of the model and the values of the parameters). More generally, this idea is widely known as the Occam's Razor principle formulated by William of Occam in 1320 in the following form: shave off all the "unneeded philosophy of the explanation" (Mitchell, 1998).

10. Define instability in the model (critical areas where small changes in inputs lead to large change in output).

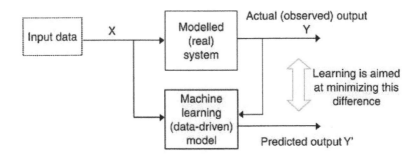

Figure 4.3: *Learning process of data-driven models (Solomatine (2005)*

11. Define uncertainty in the model (critical areas and ranges in the data set where the model produces low confidence predictions or insights).

The training process is the minimization of the difference between the measured and predicted results, through the update of model parameters. It would appear that many researchers give a lot of attention to the generation of the model used, and being more specific, the training process (Figure 4.3). However, the data-driven model is as good as the data used to build it, so the issues of data preparation and optimal partitioning are as important, or perhaps even more important.

4.4   Preparing data and building a model

Data-driven modelling assumes that data is split in a number of sub-sets to aid the process of building and assessing the final performance of the model.

Training data set: Data used to update the internal parameters of the models; it is analogous to calibration data set in process-based modelling.

Validation (cross-validation) data set: Data required for the evaluation of the model performance after training. This data is unseen by the model, but is used to tune either the model structure (e.g. node in a network), or to aid the early stop of the adaptation process.

Testing (verification) data set: Data imitating the process of model operation. The performance assessment of a model, before it is used, requires data that have not been used by the model, either explicitly or implicitly.

Models training on (even slightly) different data sets will have different performance, so data has to be partitioning in a way leading to the optimal model.

Cross validation imitates the test set and is used during training to judge the quality of the model at a particular phase of training. One of a popular ways to perform cross-validation is to creating a number of different possible splits of data into training and cross-validation subsets. Ten-fold cross validation is one of the most commonly used techniques and is applied in most of the ANN models developed in this thesis. In it, ten pairs of training – cross-validation sets are created, by taking 9/10 of data for training, and the rest – for cross-validation; ten different models are trained. If performance of all the models on the cross-validation set is similar, it is an indication that the model building is on the right track, and either the best model, or an ensemble of the ten models is used. Note that this procedure does not guarantee the model optimality.

Four steps in model building can be distinguished (Figure 4.4).

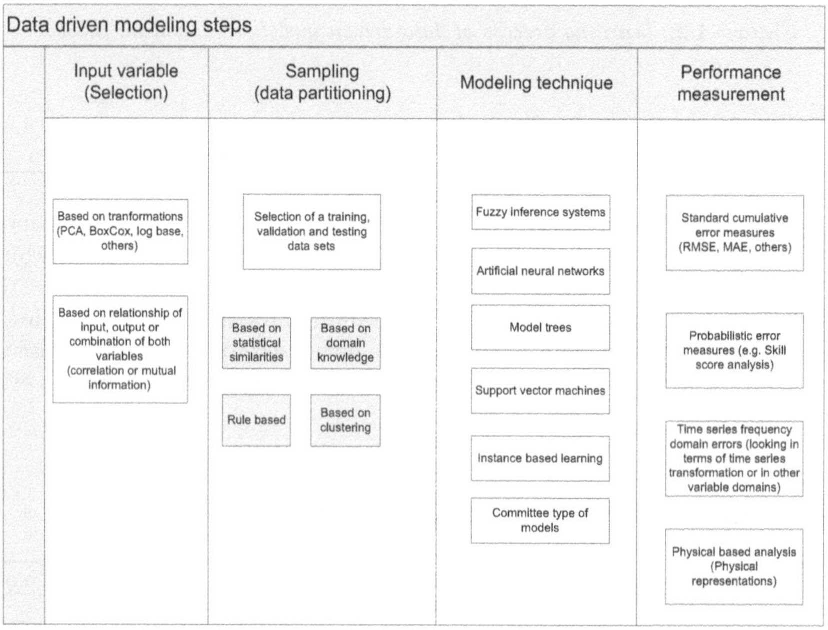

Figure 4.4: *Steps in building a data-driven model*

Step 1. Selection of input variables, data transformation

As it can be seen in Figure 4.4, the input variable selection can make use of several methods. The most common technique is based on the correlation analysis. Another method is using average mutual information. A more detailed discussion on the procedures used here can be found in the publications presented by Bowden et al. (2005a,b).

One of the important steps in DDM is data transformation (pre-processing and post processing of data) (Pyle, 1999). For some of the methods, like MLP ANNs data transformation (normalization) is almost a must. Other methods benefit from it because it leads to improved performance.

Step 2. Data partitioning

As mentioned before, to be able to judge the performance of the model it is necessary to have unseen data. However, sometimes it is argued that this could lead to a weaker model; since not all the information available was used in training.

One may find enough publications where researches do not give attention to validation their results; see the analysis of typical examples of this in (Solomatine and Ostfeld, 2008). Often an objective is to just demonstrate particular cases of a modelling practice and not to generalize the model results. In this chapter a comparison of the differences in performance for the training, validation and testing is made. It should also mentioned that the lack of attention to testing is even more common in the use of physically based models where all the data is used to calibrate a model and testing is practically ignored.

Step 3. Training the model

The third step shown in Figure 4.4 is the modelling process per se, which in principle is an optimization process to choose the model parameters and the structure to fit the model output to the measured (target) values. There are a number of single and muti-objective optimization techniques available for such purpose. Multiple criteria may lead to so-called multi-objective learning.

It is not possible to state that "one technique will always provide a better solution than any other". Some algorithms inherently contain the random components that lead to a problem of non-reproducibility: for example, ANN training is based on the initial randomization of weights. This problem is addressed later in this chapter.

Step 4. Validation and testing

After the model is optimized, and, as such, probably selected after a number of validation iterations, the model error on testing data can be used to asses the resulting model performance. Additionally, such model characteristics as uncertainty and robustness can be tested as well. Commonly, visual comparison and a benchmark, or reference, model is used to asses the quality of the model.

From the cyclic representation of data-driven processes interactions presented in Figure 4.5, it is possible to see the interaction of the modelling steps, and how they contribute to the other steps of the overall modelling process. In the following sections more details on some issues considered important for the modelling process; necessary for having an idea of the possible performance changes with different decision in each steps.

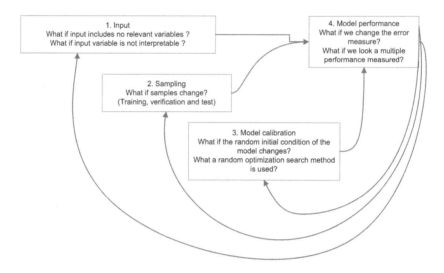

Figure 4.5: *Links between question that need to be answered when building a data-driven model*

4.5   The problem of input variables selection

Data-driven rainfall-runoff models are commonly formulated using the major variables (precipitation, evapotranspiration, discharge, and sometimes temperature). The model normally uses their measurements to increase the available information about the system. The notion of the lag (travel) time plays an important role here. The general form of a typical data-driven rainfall-runoff model (involving only effective rainfall and discharge) is presented in the previous chapter by Equation 3.1.

In DMM selection of input variables is of great importance, and this process is quite different to the one normally used in process based modelling. Having a large volume of measurements does not guarantee high accuracy of a model. Even knowing what input variables are driving the output in physical terms is not enough to build accurate models. In most cases it is necessary to identify "when" and "to what extent" these inputs contributed to the model output. Since in DDM the information about the modelled phenomena is not always available, most methods rely on the statistical and information theory-based methods to determine the appropriate input variables for data-driven models.

Normally, the input selection process starts with all the knowledge (data) about the process that will be modelled, and later the selection space is narrowed based on the subsequent more detailed analysis. In contrast to process-based modelling, DMM allows for inclusion of any variable (or their combination), even of those that are not necessarily forcing the phenomenon (discharge) directly. For example, rainfall-runoff models may use the past discharges to forecast the current or the future ones, however, they are not the actual immediate trigger of flood situations.

To take into account the past information about the modelled phenomenon, lagged variables may be used. In the context of hydrological forecasting, these are precipitation and discharges. Two mostly widely applied analysis techniques used to select the appropriate input variables and their lags are correlation analysis and the average mutual information (AMI) analysis. The correlation analysis reflects only linear relationships so when processes are highly non-linear, AMI would be a better choice.

In the studies presented here, the lag between two time series (lag inside one series) is defined as the number of time steps by which a time series is shifted relative to a reference time (when cross-correlation is studied), or relative to the corresponding time values of the same series (when auto-correlation is studied). This procedure is shown in Figure 4.6. After the shift in time is done, a new vector with past data is created.

The last component in it is the recorded value (taget) of discharge to be forecasted (for the past it is known); the rest of variables are inputs. In this process the records with no past information are removed. It is necessary to make an assumption on the continuity of the time series, so it is assumed that the vector has a number of past phenomena that covers more than the response time of any precipitation phenomena. Figure 4.6 highlights the sample for Jan 7/1988 to become the first one in the data matrix: it has the previous 4 days of precipitation in the analysis (till 3 Jan/1988).

Precipitation					Discharge
'03-Jan-1988' 7.1					23.78
'04-Jan-1988' 5.9	7.1				42.06
'05-Jan-1988' 3.6	5.9	7.1			51.97
'06-Jan-1988' 4.8	3.6	5.9	7.1		58.24
'07-Jan-1988' 1.2	4.8	3.6	5.9	7.1	59.79
'08-Jan-1988' 0.1	1.2	4.8	3.6	5.9	57.76
'09-Jan-1988' 0.5	0.1	1.2	4.8	3.6	52.22
'10-Jan-1988' 5.3	0.5	0.1	1.2	4.8	46.12
'11-Jan-1988' 0	5.3	0.5	0.1	1.2	41.93
	0	5.3	0.5	0.1	
		0	5.3	0.5	
			0	5.3	

Figure 4.6: *Building the data matrix based on the lagged variables*

Identification of the delay can be based on the analysis of the physical process at question, on the analysis of relatedness of the lagged input variable to the output, or on both. For example, for the case studies presented in this chapter the 10 days lag is used: it can be shown that the precipitation observed earlier than 10 days does not have significant relation on the current and future flows.

Note however, that since the average precipitation is considered, it is often not enough to include only one or two lagged variables into the model. Precipitation events taking place close to the discharge measurement point would lead to an increase of discharge with a lag which could be much smaller than the average one, and events far from this point would have larger lag.

4.5.1   Inputs selection based on correlation analysis

The correlation coefficient is commonly used to determine mathematical linear relations between the two samples of random variables, or time series; in case of building a rainfall-runoff model the variables are lagged precipitation and discharges (Equation 4.1):

$$\rho_{xy} = \frac{n \sum_{i=1}^{n} x_i y_i - \sum_{i=1}^{n} x_i \sum_{i=1}^{n} y_i}{\sqrt{n \sum_{i=1}^{n} x_i^2 - \left(\sum_{i=1}^{n} x\right)^2} \sqrt{n \sum_{i=1}^{n} y_i^2 - \left(\sum_{i=1}^{n} y\right)^2}} \tag{4.1}$$

Where x_i and y_i are samples of the precipitation and discharge respectively, n is the number of sample available in the time series. $\bar{x}$ and $\bar{y}$ are the sample means, and is the correlation coefficient. If x and y are lagged discharged and actual discharge, it is said that is an auto-correlation analysis (Figure 4.7).

The problem is to determine the lag time leading to high correlation. Since the correlation coefficient can be misleading if not enough data is used, or there could be events with the different correlation structure, it could be useful to employ two other analysis techniques: (a) to analyse the variation of the correlation coefficient with different sizes of data sets, and (b) to perform the correlation analysis separately for certain events, for example, with the discharge in a particular range.

Analysis of correlation for data sets of varying size

It is indeed a concern that the correlation between precipitation phenomena and discharge may change considerably with the size of the available data set. To find out the influence of the amount of data on correlation, we calculate the cross-correlation values between actual or lagged precipitation and actual discharge for different sizes of the data set. The size of the data set starts from one record corresponding to the first date available, and then progressively adding one record (one day) and calculating the correlation. This can be interpreted as a time window size of n samples. The size of the window is iteratively increased by adding the next day to the calculation. This procedure is continued until the whole data set is included.

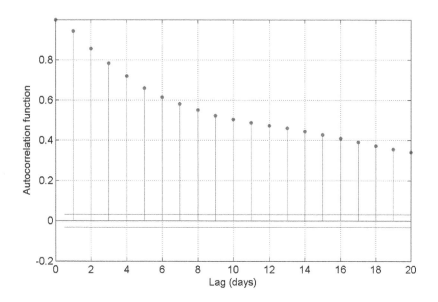

Figure 4.7: *Autocorrelation for the discharge at Ourthe*

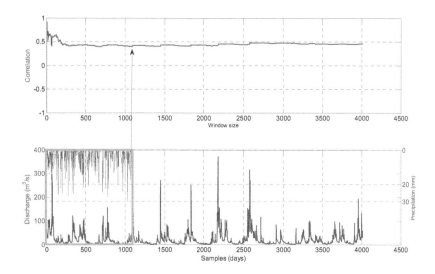

Figure 4.8: *Correlation of the lagged precipitation (3 days) and discharge for diffe-rent window sizes*

The variation of the correlation due to the changes in the amount of data, with different time windows is shown in Figure 4.8. The window sizes taken started from 2 samples till 4014 samples. The upper part of the figure shows from left to right the correlation obtained with a window of the size of the number of samples mentioned in the x axis. It is possible to see that the correlation between the precipitation and discharge stops changing after the window becomes wider than 300 days. Accordingly, the response of the catchment (correlation between precipitation and discharge), can be said to be characterized well with at least 1 year of data.

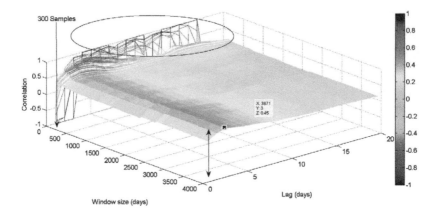

Figure 4.9: *Correlation for different window size and for different lags of the precipitation time series*

For the comprehensive analysis of multiple lags, 3D visualization is presented in Figure 4.9. The value of correlation shown is plotted with a variable lagged (from 0 to 20). From this view all the lagged time series have similar shape to that in Figure 4.8, with the same threshold of 300 samples (encircled region on Figure 4.9). An interesting and encouraging fact is that the plots have the common peak of the correlation coefficient for the lag of 3 days.

Analysis of correlation for different rainfall-runoff events

The physics of the rainfall-runoff process is such that precipitation events of different magnitude and location have different response times. Correlation analysis allows for studying this effect since the corresponding correlation coefficients will have different values for different events. Analysis of this phenomenon is important for determining lags and building accurate DDMs.

Figure 4.10, shows the correlation between the lagged precipitation (10 steps) and the actual discharge, for its ranges above a variable threshold (y axis). Zero represents the use of all the data available. This results shows that the correlation is similar for events that included discharge under $180m^3/s$; a 3 days peak correlation. For events with discharges above $180m^3/s$, there is no clear dominating correlation coefficient; some of these values show high

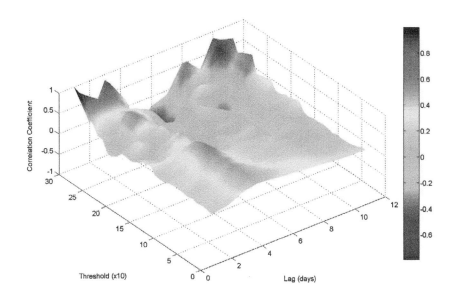

Figure 4.10: *Correlation for different threshold and different lags*

correlation at 2 or even 1 day. This is expected since high flows in this basin are mainly driven by intense precipitation with different fast flows.

Since the number of events that have high discharge is quite small, it is important to quantify the response times. Figure 4.11 presents the precipitation lags corresponding to the maximum correlation with discharge and the corresponding number of samples used. It is possible to see that on the threshold $180m^3/s$ the amount of samples is almost 30.

4.5.2   Selection based on Average Mutual Information (AMI)

Average mutual information has shown to be very useful in selecting inputs for data-driven models (Abebe and Price, 2003; Bowden et al., 2005a,b; Solomatine and Dulal, 2003).

The AMI between the two measurements x_i and y_j drawn from sets X and Y is defined by:

$$I_{XY} = \sum_{x_i y_j} P_{XY}(x_i, y_j) \log_2 \left[\frac{P_{XY}(x_i, y_j)}{P_X(x_i) P_Y(y_j)} \right] \tag{4.2}$$

$$P_{XY}(x_i, y_j) = \iint\limits_{XY} f(x, y) \mathrm{d}x \mathrm{d}y \tag{4.3}$$

where $P_{XY}(x_i, y_j)$ is the joint probability density for measurements X and

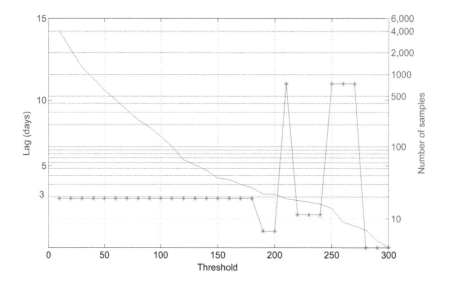

Figure 4.11: *Precipitation lags corresponding to the maximum correlation with discharge and the number of samples used.*

Y resulting in values x and y, $P_X(x_i)$ and $P_Y(y_i)$ are the individual probability densities for the measurements of x_i and y_i. If the measurements of a value from X resulting in x_i is completely independent of the measurement of a value from Y resulting in y_j then the average mutual information I_{XY} is zero. For the considered hydrological modelling problem X would typically stand for discharge, and Y the lagged precipitation.

Analysis on the basis of AMI seems to be quite straightforward, but AMI is sensitive to the selection of the bin size, and was found to be different for different flow events. The analysis of these aspects is presented below.

AMI sensitivity to the bin size

In this analysis probability distributions of observations are generated based on the bins with the sizes varying from 1 to 50 (so every time bin size is different). AMI for discharge and the lagged precipitation was calculated for different lag times varying from 1 to 20. Figure 4.12 presents the results.

Although AMI changes with the bin size change, its peak value always correspond to the same lag value. The AMI value is not measurable on a fixed scale and therefore it does not matter which peak points to lag to choose (Abebe and Price, 2003). The results are consistent in shape and AMI peaks point to almost the same lag values as the plots of the correlation coefficients do.

Bin size of 10 shows the first peak of the AMI at the lag of 3 days. If a bin

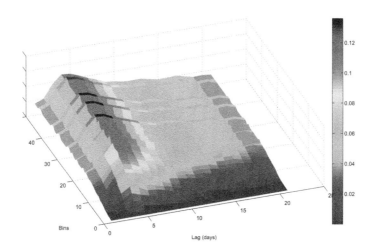

Figure 4.12: *AMI for different bin sizes and lags*

of 30 is used, the AMI value for 3 days is almost the same as for 4 days. In such cases it could be wise to consider for inclusion into the model the precipitation variables with the lag 3 and 4.

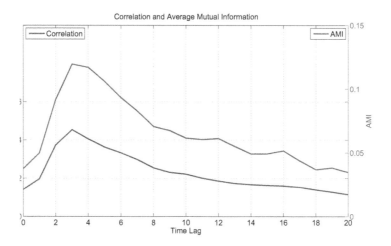

Figure 4.13: *AMI calculated for a bin size of 30*

AMI for different flow events

The procedure to form events is the same as used in correlation analysis. AMI was calculated for the data subsets with the discharges above a certain threshold, varying from 10 to 300m m^3/s.

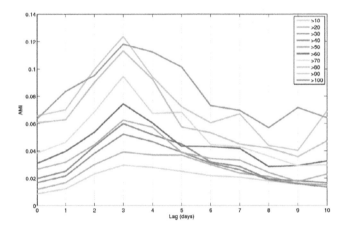

(a) *AMI for threshold from 10-100*

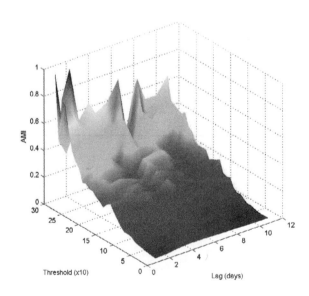

(b) *Ranges from 0-300 thresholds*

Figure 4.14: *AMI results for precipitaot be different tion and discharge (10 days lags)*

Figure 4.14a and 4.14b, show that the AMI is a more sensitive measure than the correlation coefficient. AMI is higher for the events with discharge between 10 and 100. Shape of the AMI plots with different lags is similar. Above 100 m^3/s the AMI does not allow for choosing the lag: in some cases the peaks correspond to one hour or two hour lag. Since most high flows typically result from the high intensity precipitation, this finding indicates that it is risky to rely only upon the average lag of 3 or 4 days found when full data set is used to calculate AMI.

4.6   Influence of data partitioning on model performance

Partitioning the data into training, validation and testing sets should typically result in sets which are "statistically similar". In practice, however, partitioning procedures may be influenced by the amount of the available data and the specifics of the problem. For example, lack of data forces some modellers to omit the cross-validation set. Due to the specifics of hydrological modelling it is often required that all the three (or two) data sets are contiguous, that is samples are sequential in time, so that the output would be represented as an interpretable hydrograph (Solomatine, 2005).

From purely statistical point of view it is recommended to have similar properties (mean, max, min and standard deviation), on the training, validation and testing sets. Although this procedure does not guarantee that the model will be trained on the optimally constructed subsets (like it is done, e.g., by Bowden et al. (2005a) with the help of GA optimization) but in practice this approach is often used, and is good enough for most practical purposes. The procedure adopted in this chapter follows two steps.

1. Generate 100 different groups of randomly generated training, cross-validation and tests sets,

2. Choose the group in which the three mentioned sets are statistically similar to a maximum extent using some appropriate measure of similarity.

For training, validation and testing 50, 16 and 34 percent of the available data (4016 records) are used, respectively. The discharge measure is used as the split variable. To compare the statistical characteristics of the three data sets it is necessary to asses the absolute difference in terms of a combination of statistical measures. Therefore, the three following measures were evaluated.

$$RADS_{tv} = \left| \frac{\sigma_{tr} - \sigma_{va}}{\sigma_{tr}} \right| \qquad (4.4)$$

$$RADS_{tt} = \left| \frac{\sigma_{tr} - \sigma_{te}}{\sigma_{tr}} \right| \qquad (4.5)$$

$$RADM_{tv} = \left| \frac{\mu_{tr} - \mu_{va}}{\mu_{tr}} \right| \qquad (4.6)$$

$$RADM_{tt} = \left| \frac{\mu_{tr} - \mu_{te}}{\mu_{tr}} \right| \qquad (4.7)$$

Where, $RADS_{tv}$ is the relative absolute difference between the standard deviation of the training and validation data set. $RADS_{tt}$ is the relative absolute difference between the standard deviation of the training and testing data set. $RADM_{tv}$, is the relative absolute difference between the mean of the training and validation data set. $RADM_{tt}$, is the relative absolute difference between the mean of the training and testing data set.

The different measures are compared in a plot with the 100 random splits (Figure 4.15). The figure shows the four relative absolute difference measures, of the 100 generated groups of the three sets (training, verification and testing). Since there is no unique or perfect data sampling that will contain identical statistical properties for all the three data sets required, multiple solutions to this can be plotted in a Pareto front in 4 dimensions (Shamseldin and O'Connor, 2001). However, here a simple visual analysis of the 100 splits in each dimension of the problem is undertaken. The criterion was to find the partitions for which the values of the four criteria would not be too much different (say, would not differ more than 10 %), This process was used to select the 12 groups from the Figure 4.15. The groups selected where '1', '15', '21', '44', '54', '56', '60', '67', '86', '9', '91', 'H'. The last group (H) of samples was constructed of the selected continuous samples of the hydrograph (e.g. training from January 3 of 1988 till 13 July 1993, validation and testing the remaining period till December 1998).

4.7   Influence of ANN weight initialization on model performance

Training of data-driven models involves the adaptation of weights or parameters in such a way that the model fits the measured data. In case of using MLP ANNs, this process starts with some initial state of the weight or parameters, and often random initialization is used. Different initial weights may lead to models with different performance. To explore this performance variance due to various initializations, 20 ANN rainfall-runoff models are set-up. In this process two situations of models are considered, rainfall-runoff model with the past discharges, and without them (denoted as RRQ and RR respectively) .

4.7.1   Models not using past discharges as inputs (RR)

Tables 4.1, 4.2 and 4.3 show the performance of 20 ANN models generated for the Ourthe river basin when the input vector does not include the past discharge. These models used the actual and previous precipitation from the past 4 days as input, and the target considered was the discharge one day ahead. The models where set up with the same number of hidden nodes (4) but with the different initial random weights. All were trained with neither

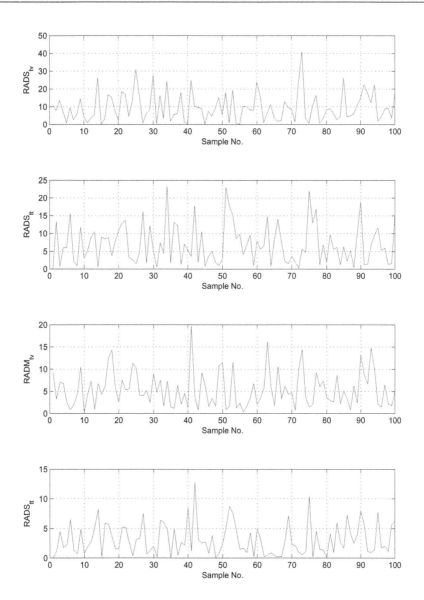

Figure 4.15: *Statistical characteristics of the 100 generated training, validation and testing sets*

optimizing the nodes, nor having feedback from the validation or testing set. The artificial neural network optimization algorithm used was the Levenberg-Marquardt (Mor, 1977) for the mean square error minimization. All the models were trained using 200 epochs and the learning rate of 0.1. These values were determined after multiple trials.

Table 4.1: *Training performance of 20 ANNs trained with different initial conditions (RR)*

	RMSE	NRMSE	CoE	Correlation	PERS	MARE	MAE
Mean	22.51	74.05	0.45	0.64	0.18	1.67	15.37
Min	21.25	69.91	0.001	-0.24	-0.49	1.59	14.81
Max	30.4	100	0.51	0.72	0.27	2.24	19.15
Std	2.2	7.23	0.12	0.21	0.18	0.16	1.05
Std/Mean	9.76%	9.76%	27.09%	33.13%	103.02%	9.38%	6.81%

From Table 4.1 it can be seen that the results are significantly different. In training the results the RMSE and NRMSE show that the bound of the standard deviation is 9.76% of the mean error value. Correlation coefficient, Nash Sutcliffe coefficient(CoE) and PERS indicate the low model performance. The PERS (index of persistence) shows a higher ratio between the standard deviation and mean value, indicating the high variance in terms of performance. From the Nash-Sutcliffe and correlation coefficients, it is possible to see directly that even the best network model in the training process has low accuracy. The RMSE measure has the low ratio between the standard deviation and the mean, being an error measure that is less sensitive to different random initial conditions. The maximum variation that may be expected in the random initialization can be considered to be the 10%. Therefore, in the experiment presented here this threshold is used to determine if a model may be better than another due to a chance.

Table 4.2: *Validation performance results of 20 ANN trained with different initial conditions (RR)*

	RMSE	NRMSE	CoE	Correlation	PERS	MARE	MAE
Mean	26.91	90.61	0.17	0.47	-0.33	1.83	16.63
Min	23.34	78.6	-0.07	-0.1	-0.71	1.71	15.74
Max	30.7	103.36	0.38	0.62	0.01	2.35	18.73
Std	2.2	7.42	0.14	0.15	0.22	0.15	.65
Std/Mean	8.19%	8.19%	7.805%	32%	-66.93%	7.97%	3.91%

From Tables 4.2 and 4.3 we can see the low differences between the means of different types errors on the validation and testing data sets. The results show that differences in the mean for RMSE and other variables are very small, however, the standard deviation of RMSE, NRMSE and PERS is twice as high. So, general conclusion is that the ANN models of RR type have quite a low performance.

Table 4.3: *Testing performance results of 20 ANN trained with different initial conditions (RR)*

	RMSE	NRMSE	CoE	Correlation	PERS	MARE	MAE
Mean	26.78	91.38	0.16	0.45	-0.43	1.74	16
Min	25.04	85.43	-0.001	0	-0.72	1.64	15.36
Max	29.33	100.07	0.27	0.53	-0.25	2.29	18.41
Std	1.2	4.08	0.075	0.11	0.13	0.16	0.67
Std/Mean	0.0447	0.0447	0.4625	0.2483	-0.2972	0.091	0.0416

4.7.2 Models using past discharges as inputs (RRQ)

Tables 4.4, 4.5 and 4.6 shows the performance of 20 ANN models generated for the Ourthe river basin. The input vector for these models included the current and the past precipitation for 3 days, and the past discharge for 3 days. The output of the models was discharge one day ahead. The other characteristics of the experiments were the same as for RR models covered in the previous sub-section.

Table 4.4: *Training performance of 20 ANNs trained with different initial conditions (RRQ)*

	RMSE	NRMSE	CoE	Correlation	PERS	MARE	MAE
Mean	4.11	13.53	0.98	0.99	0.97	0.12	1.89
Min	3.45	11.36	0.94	0.97	0.91	0.09	1.59
Max	7.54	24.79	0.99	0.99	0.98	0.33	3.8
Std	1	3.3	0.01	0.01	0.02	0.06	0.56
Std/Mean	0.2436	0.2436	0.012	0.0061	0.0181	0.5291	0.2938

From Table 4.4 and one can see that the performance is considerably higher than that of RR ANN models that do not use past discharges as input. This can be explained by the fact that the system does not change significantly from one time step to the next, so the auto-correlation information of the discharge have high impact on the result. The RMSE and NRMSE have high variability. Since the model is more accurate, the ratio between the standard deviation of the measures and the mean is higher; this is a consequence of the lower mean error and variability. With the 20 models generated, the standard deviation of RMSE was only 1 m^3/s.

For the verification and testing it would appear that in testing the error is more variable than in training; difference is 48% for RMSE, 5% for correlation and 18% for PERS (Table 4.6). Based on this, it is confirmed that a conclusive comparison of a model and the random initialization can not be bound clearly to this ratio or relative ranges mentioned.

The difference in performance between the training and validation and testing is high. This result can be attributed to either lack of generalization in the ANN models, or to overfitting. The overfitting is less probable since the number

Table 4.5: *Validation performance results of 20 ANN trained with different initial conditions (RRQ)*

	RMSE	NRMSE	CoE	Correlation	PERS	MARE	MAE
Mean	9.35	31.49	0.88	0.94	0.81	0.15	2.5
Min	5.53	18.61	0.63	0.85	0.41	0.11	2.03
Max	17.95	60.43	0.97	0.98	0.94	0.35	3.81
Std	4.54	15.27	0.11	0.05	0.18	0.06	0.48
Mean/Std	0.48	0.48	0.13	0.05	0.23	0.43	0.19

Table 4.6: *Testing performance results of 20 ANN trained with different initial conditions (RRQ)*

	RMSE	NRMSE	CoE	Correlation	PERS	MARE	MAE
Mean	8.03	27.38	0.91	0.96	0.85	0.12	2.22
Min	4.55	15.53	0.66	0.83	0.41	0.09	1.81
Max	17.21	58.74	0.98	0.99	0.96	0.32	3.75
Std	3.63	12.4	0.09	0.04	0.15	0.06	0.51
Mean/Std	0.45	0.45	0.1	0.05	0.18	0.48	0.23

of samples and epochs is reasonably high and fits with empirical formulations (Wang, 2006).

The variation in performance due to the differences in the weights initialization is relatively low if compared to that due to the input variable selection (From an RMSE of 26.78 to a value of 8.03 and from CoE of 0.16 to 0.91 in the testing set).

4.8   Various measures of model error

The model error can be measured using measures. In this thesis the error analysis is meant to answer two questions: 1) what is a good or acceptable model? 2) which model outperforms the other ones?. The information obtained from overall error measurements may be subjective. If it is used to compare different models it may not be clear since measures are highly relative as we saw in the previous section. In case of RRQ, CoE has a very small standard deviation (0.09). Therefore, if this measure is used alone, any of the 20 ANN models could be selected.

Commonly most modelling techniques use in calibration (training) only one single objective function. However, other measures, often based on expert judgement, are used to evaluate the model as well. However, one would think on the idea of including the entire possible objectives in one single model (multi-objective), but the final criteria on what to select will depend on a number of models in multiple dimensions. Although the multi-objectives approaches are being explored as well, the concepts contemplated in this thesis are based on single objective models.

To make a decision of selecting a model one may consider multiple error measures. Table 4.7 shows the lowest and highest performance measures of the best ANNs of type RRQ (on validation data). The correlation coefficient and the PERS seem to be less sensitive to the random initialization in the ANN models; in this sense no clear selection can be done. However, we can see the PERS and Nash-Sutcliffe coefficient have values high enough for accepting any of the models. If we extend the analysis to measures like the RMSE, MAE and NRMSE, the set of models to select from can be reduced to only models 5 and 7.

Table 4.7: *Validation performance results of the best ANN RRQ trained with different initial weights*

Network No.	RMSE	NRMSE	CoE	Correlation	PERS	MARE	MAE
2	6.04	20.33	0.96	0.98	0.93	0.11	2.09
3	12.55	42.24	0.82	0.91	0.71	0.11	2.52
4	**5.53**	**18.61**	**0.97**	**0.98**	**0.94**	**0.11**	**2.03**
5	5.76	19.38	0.96	0.98	0.94	0.11	2.11
7	**5.54**	**18.64**	**0.97**	**0.98**	**0.94**	**0.13**	**2.23**
9	5.66	19.06	0.96	0.98	0.94	0.11	2.11
10	5.87	19.77	0.96	0.98	0.94	0.11	2.18
14	6.36	21.41	0.95	0.98	0.93	0.11	2.2
18	5.86	19.73	0.96	0.98	0.94	0.11	2.14
20	6.08	20.48	0.96	0.98	0.93	0.13	2.3

Note, that ANN training algorithms use gradient descent techniques based on the fixed objective function being the mean square error (equivalent to RMSE). In this work we use also NRMSE, along with the correlation coefficient, complemented by the visual inspection of the resulting hydrograph.

Another important issue to take into account is that it seems that due to the random weight initialization, with an RRQ type of model, the performance of an individual model could be around 40% worse than the mean. Further statistical analysis of the probability distribution shows that 56% of the models have RMSE around ± 10% of the mode of the samples. From the probability density function we can see the best fit obtained (log normal distribution is used 4.16a).

4.9 Comparing the various types of models

The variability of the performance of different data-driven types of models is considered here. Data-driven models have shown to be an accurate alternative for building rainfall-runoff models but it is impossible to recommend one model type that will be best in all cases. In this section performances of several models are compared: MLP ANNs (ASCE (2000a)), model trees (Solomatine and Dulal (2003); Witten and Frank (2000)), instance based learning (Solomatine et al.) and support vector machines (Bray and Han (2004); Cortes and Vapnik (1995)).

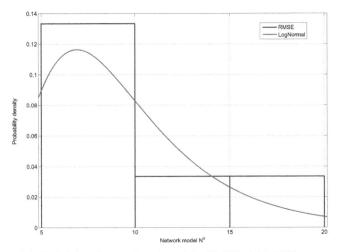

(a) *Probability density function of the RMSE of 20 ANN runs*

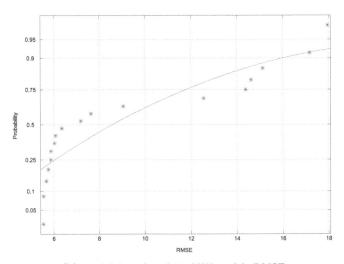

(b) *Probability plot of 20 ANN models RMSE*

Figure 4.16: *ANN RMSE probability distribution*

These modelling techniques are briefly presented in the Appendix B. All models were built with the same input variables using the 12 data sets sampled in the section on data partitioning. The optimization of the model structure was performed as follows:

- Artificial Neural Network: MLP networks are trained by the Levenberg-Marquardt algorithm (Mor, 1977). For this a learning rate (0.1) with 200 epochs are used. The transfer function for the hidden nodes is a sigmoid, for the output a linear transfer function is used. Data is normalized to the [0, 1] interval. A sigmoid transfer function is used in hidden nodes and a linear transfer in the output node. The structure of the neural network is mainly optimized by finding the right number of hidden nodes. For this purpose, 30 neural networks with the number of nodes varying from 1 to 30 were built. The selection of the best network was based on its error on the validation set, and the final performance was assessed with the test set.

- Instance based learning (k-nearest neighbours method): The number of neighbours used was determined by exhaustive optimization. From 1 to 50 neighbours where analyzed, and the minimum error in the validation was used to make the selection of the model.

- Model trees (piece-wise linear regression models, M5' (M5P) alogrithm): The main method for tuning the performance and improving the generalization in a model tree is pruning. In this case, the algorithm used is based on Witten and Frank (2000), which uses the minimum number of instances per leaf. A higher this number is, the less leaves are created. Models were built with the number of instances per leaf ranging from 3 to 30. The decision on the best model performance was based on the model error on validation set.

- Support vector machine: The optimization of the support vector was based on the two criteria. One was the kernel function, and three of them were tested: linear, polynomial, and radial-basis functions (RBF). The second criteria for the optimization are the parameters C and Gamma, which are found through a grid search process with logarithmic steps.

Figure 4.17 shows the RMSE validation results of 6 data-driven models using 12 data sets. The ANN model had the lowest error on almost all validation datasets. The SVM models with the linear kernel follow the ANN in performance. They are followed by the instance based learning and the M5P model tree algorithms which have very similar performance. SVM with the RBF function kernel, and especially with the polynomial kernel are not performing well.

For the testing data set, it would appear that the performance of the ANN models is even better. The other models deteriorate a little, it is the case of the SVM with linear kernel, which is on par with model trees and instance based learning.

The computational time used in the optimization of the instance-based and the model tree is the lowest (25 to 45 minutes), and the SVM training and optimization needed most of the time (6 to 8 hours), and ANN model was in this respect average with 2 hours needed. These considerations could be important for the efficiency of research, but not for the decision about which model to finally choose to be used in operation: the training and optimization times are indeed quite low for all models. It is important to highlight that the

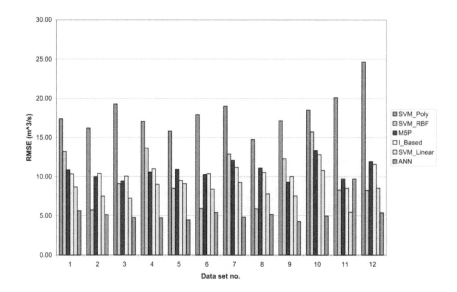

Figure 4.17: *Validation errors of six types of models trained on 12 different training sets*

actual HBV conceptual model with expert calibration reaches an RMSE value of 6.26 and a CoE of 0.91.

4.10 Discussion and conclusions

Important procedures related to data partitioning and the variability of data-driven models have been discussed and explored on an example of building a rainfall-runoff model. The sensitivity of the inputs due to data availability and with respect to different types of flow events was analyzed. Six different model types were evaluated with 12 different data-sets.

The following are some of the conclusions drawn from the experiments in this chapter.

The input selection process (including the selection of lags) using correlation and average mutual information brings similar results. Correlation analysis allows for a clearer picture of the dominant lags, while the AMI provides an interval of the lags to use. The correlation has the disadvantage that is highly sensitive to the available data. The tests performed in this chapter are suggested to be used for exploring such sensitivity.

The precipitation information as input to rainfall-runoff models oriented at forecasting may not be sufficient, and may require the use of past discharge(s) as inputs to the model. It is known from experience of different researches that the difference in performance of using as only input lagged discharge is quite

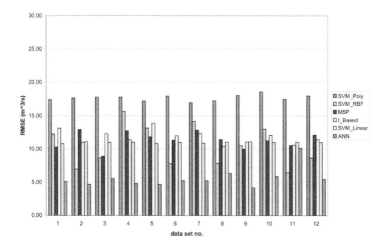

Figure 4.18: *Testing errors of six types of models trained on 12 different training sets*

similar to the one of using lagged precipitation with the lagged discharge.

To study the influence of weights initialization on the ANN model performance, 20 ANN models were trained based on different initial vectors of weights. Most of their RMSE errors were found to be close to the mode (around 10% lower or higher than the mean). There were a number of models that went out of the bounds of the standard deviation of RMSE. We can conclude that the variability of the ANN models performance due to the differences in the weights initialization needs to be always considered. For this reason the procedure of optimizing network model structures and the 10 fold cross validation is used in the following chapters. The maximum RMSE error is less that the differences in performance of the models with different structures (using different input variables). In further chapters the value of 10% will be used as a reference to compare with other models, although the proposed value is less than the one found in the experiments described in this chapter, we expect to obtain a representative value when the results of structure optimization (30 ANN runs) and 10 fold cross validation provide the best ANN.

From these results, it has been concluded that the ANN model would be the best one for building data-driven models. The procedures used in this chapter will be used for all the data-driven models developed in this thesis.

TIME AND PROCESS BASED MODULARIZATION IN LUMPED RAINFALL–RUNOFF MODELLING

Chapters 2 and 3 introduced the concept of modularization, and Chapter 4 has shown the different data-driven modelling process. This chapter explores the application of the modular models presented in Chapter 3.

Attempts to improve data-driven forecasting models relate, to a large extent, to the recognized problems of their physical interpretation. The present chapter deals with the problem of incorporating hydrological knowledge into the modelling process through the use of a modular architecture that takes into account the existence of various flow regimes. Three different data partitioning schemes are employed: automatic classification based on clustering, temporal segmentation of the hydrograph based on an adapted baseflow separation technique, and an optimized baseflow separation filter. This chapter discusses the results obtained on the application of the mentioned models to three case studies (Corzo and Solomatine, 2007a).

5.1 Introduction

Modular modelling procedures contemplate partitioning of data to represent a specific part of a problem. In this sense the problem to be solved needs to be divided in smaller units that will be represented by independent models. This approach has the disadvantage that the separation of a process in a hydrological system is sometimes not possible. In studying the rainfall-runoff relationship, the processes are analyzed as events, and normally measured information from forcing variables and stream flow are available(Price, 2000). Hydrology of most processes is quite well understood and modelled by various types of models. In this sense the formulations used for modular modelling include the traditional formulations encapsulated in hydrological models.

In this chapter the application of the schemes presented in Chapter 3 are applied to three case studies using lumped conceptual hydrological models. The modular models used here are:

- MM1 (Section 3.1, Figure 3.3): Modularization is based on clustering. Two clusters are identified within the training data set and an ANN model is built for each cluster. On the basis of clustered data, a classifier is trained, and it is used in operation to attribute data to a particular model.

- MM2 (Section 3.2, Figure 3.4): Modularization is based on the sub-process identification which is done by the hydrograph (graphical) analysis of baseflow. Two ANN models are built: the model of baseflow, and the model of the total runoff. In operation the hydrograph analysis is replicated by a classifier (Model tree).

- MM3 (Section 3.3, Figure 3.5): Modularization uses time-based partitioning of flow into two components (which corresponds to some sub-processes which are however not identified explicitly), an ANN model that forecasts both parts of the flow (baseflow and direct runoff) are trained.

For comparison, a data-driven single model has been also developed on the basis of the whole data set; referred to as global (GM), Chapter 4. An MLP ANN was used, and, for the sake of preserving the consistency between the models all models have the same structure.

5.2 Catchment descriptions

Three basins, Bagmati in Nepal (B1), Sieve in Italy (B2) (Brath et al., 2002; Solomatine and Dulal, 2003), and Brue in the UK (B3) (Moore, 2002) were considered as case studies (Table 5.1 and Figure 5.1, 5.2, 5.3). The size, location and other characteristics of the basins are significantly different, and this allowed for validating the presented modelling approach under different spatial and temporal forecasting conditions. The detailed hydro-geological description of these three basins can be found in the papers mentioned above.

Figure 5.1 shows the hydrograph of the Bagmati basin in Nepal; the basin response time is approximately one day. As it can be seen the low flow periods have regular flows under $60 m^3/s$. However, during the rainy season this could turn into flood situation with the flows above 2500 m^3/s.

The hydrograph presented in Figure 5.2 for the Sieve basin, shows a clear difference in response for dry and wet events. The response time of the basin is around six hours.

Figure 5.3 show a fast response of a small basin. The response time of this basin is quite variable; however, correlation analysis shows 10 hours in average. The basin does seem to have a highly different response to precipitation in winter than in summer.

Table 5.1: *General hydrological characteristics of the three basins.*

Basin name	B1 (Bagmati)	B2 (Sieve)	B3 (Brue)
Topography	Mixed	Mountain	Gently slopes
River length (km)	170	56	20
Basin Area (km^2)	3500	836	135.2
Daily data set from	01/01/88	01/12/59	01/09/93
To	01/12/95	01/02/60	30/08/95
Training start	01/01/88	13/12/59 19:00	01/09/93 00:00
Training end	22/06/93	28/02/60 19:00	01/08/94 00:00
Number of samples Training		28/02/60 19:00	01/08/94 00:00
Verification start	23/06/93	01/12/59 07:00	01/09/94 00:00
Verification end	31/12/95	13/12/59 18:00	01/08/95 00:00
Location	Nepal	Italy	England
Time step	1 day	1h	1h

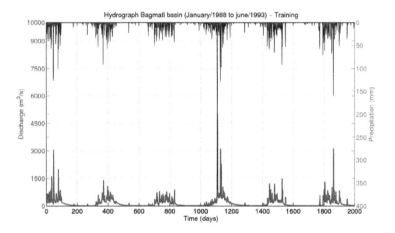

Figure 5.1: *Hydrograph built from daily time series of the Bagmati catchment from January/1988 to June/1993*

5.3   Input variable selection

The choice of input variables for global and modular ANN models was based on correlation and mutual information analysis between the input and output variables, as described in Chapter 4. The variables chosen for the artificial neural network (ANN) models are shown in Table 5.2.

Figures 5.4a and 5.4b, present the autocorrelation and cross-correlation results for the Sieve river basin. The autocorrelation till 6 hours is still higher than the maximum cross correlation. On the other hand, the peak both of the maximum correlation and the mutual information is observed at 6 hours. Along with this analysis several models with different combinations of varia-

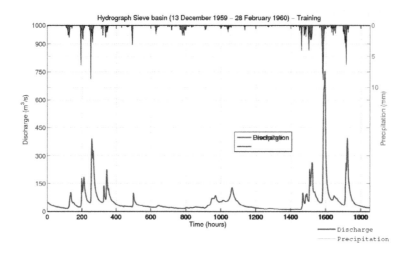

Figure 5.2: *Hydrograph built from hourly time series of the Sieve catchment from Decembre/1959 to Febraury/1960*

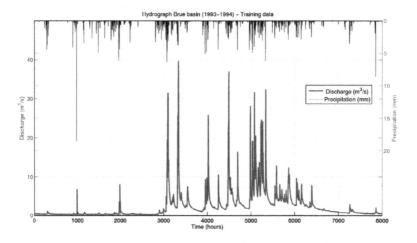

Figure 5.3: *Hydrograph built from hourly time series of the Brue basin from September/1993 to August/1994.*

bles have been run as well. As was already mentioned in Chapter 4, all models that did not include discharges as inputs had low performance. In Chapter 4 it was shown that the difference in average performance for models with and without past precipitation is very small for one time step, however, for accurate forecasting in flood situations precipitation has to be included as input. This procedure was common in all case studies and was the basis for the input selec-

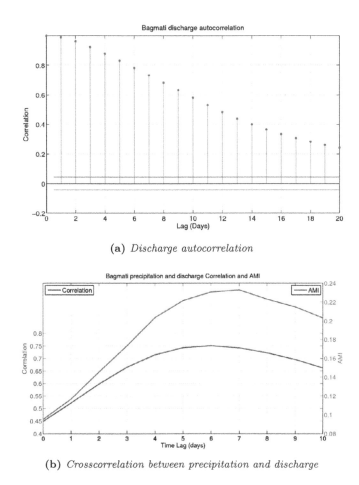

(a) *Discharge autocorrelation*

(b) *Crosscorrelation between precipitation and discharge*

Figure 5.4: *Autocorrelation of lagged discharge and cross-correlation with precipitation time series for the Sieve river basin*

tion (Table 5.2). The other basins were evaluated using the same procedures to select inputs. For extended forecast horizons only basins B2 and B3 were considered. The model handling baseflow did not include precipitation as an input.

5.4   Comparison to benchmark models

A number of conceptual hydrological models were used as benchmark models. The best conceptual model from the results obtained by Shrestha and Solomatine (2008); Shrestha (2003); Solomatine and Dulal (2003); Solomatine et al., were used for each basin. Table 5.3 presents the results of several models built

Table 5.2: *Inputs variables used for ANN forecast models*

	Basin name		
Forecast	Bagmati (B1)	Sieve (B2)	Brue (B3)
$Q_{t+1}=$	$f(P_t, P_{t-1}, P_{t-2},$ $Q_{t-1}, Q_t)$	$f(P_t, P_{t-3},$ $Q_{t-3}, Q_{t-2}, Q_{t-1}, Q_t)$	$f(P_{t-10}, P_{t-9},$ $Q_{t-1}, Q_{t-2}, Q_{t-3}, Q_t)$
$Q_{t+2}=$			$f(P_{t-8}, P_{t-9}, P_{t-7}, P_{t-6},$ $Q_{t-1}, Q_{t-2}, Q_{t-3}, Q_t)$
$Q_{t+3}=$		$f(P_t, P_{t-3}, Q_{t-3}, Q_t)$	$f(P_{t-6},$ $Q_{t-3}, Q_{t-2}, Q_{t-3}, Q_t)$
$Q_{t+6}=$		$f(P_t, Q_t)$	$f(P_{t-3},$ $Q_{t-3}, Q_{t-2}, Q_{t-3}, Q_t)$

for each of the basins mentioned. In the models for basin B1 (Bagmati) only the forcing variables (e.g. precipitation, temperature and other) were used as inputs, and the set of input variables was built following the exhaustive performance-based optimization process. In the other basins mentioned the data-driven models did include discharge input. The range of time series data used for verification is the same as the one used in this thesis. For basin B1 the best conceptual model had RMSE of $132 m^3/s$. At basin B2 there was not enough information available to build the model; temperature was not available and therefore the models had a very low performance. For basin B3 the HBV model had an average RMSE of $0.97\ m^3/s$. These values will be used further as a reference for comparison.

5.5   Modelling process

The basic structure of the ANN models was the same for all catchments: a three-layer MLP, with a tangent transfer function in the hidden layer and a linear transfer function in the output layer. The statistical parameters and distribution of the training and test data set were verified. The training was performed using the Levenberg-Marquardt algorithm; termination was based on reaching the maximum number of epochs (150) or the change in mean square error dropping below 0.0001.

For the initial clustering in MM1 we used the k-means algorithm. Classifiers in splitting models for MM1 and MM2 used RBF ANN and the Fisher discriminant method, respectively. These were selected based on their performance. Regression trees were also tested to serve as classifiers but they were less accurate.

For the ANN MLP and RBF models, the MATLAB Neural Network toolbox was used. The Fisher discriminant algorithm was based on the MATLAB Statistical Toolbox. Optimization of the MM3 model was based on the Genetic and Direct Search Toolbox developed for MATLAB. A Pentium 4 3.2 GHz PC was used.

Each of the modular models had different number of hidden nodes according

Table 5.3: *Conceptual and data-driven models performance in terms of RMSE (m^3/s)*

Basin name	B1 (Bagmati)	B2 (Sieve)	B3 (Brue)
ANN model	163.14	3.612^a , 12.55^b , 21.55^c	x^d
Tank model	179.15	x^d	x^d
HBV Model	x^d	x^d	0.97
Nam model	132.81	x^d	x^d
ADM	177.01	x^d	x^d
Model Tree (M5P)	153.58	5.17^a, 11.35^b , 19.40^c	0.45^b

[a]forecast 1 hour
[b]forecast 3 hours
[c]forecast 6 hours
[d]x not available or the period of testing was different.

to the catchment and modular scheme used. The inputs for the modular models were the same as in the global model; however, in the MM3 the baseflow component by definition should not include the precipitation as input. Analysis of the convergence of the MLP training in different trials showed that 150 iterations were sufficient in all cases. For optimization of the MM3 scheme, two algorithms, GA and Pattern Search (Abramson et al., 2004) were used in all experiments. It appeared, however, that the GA was too slow, especially for basins B2 and B3 where it did not show any sign of convergence even after 24 hours of computation. The experience with the GA cannot be characterized as positive; however, this could be attributed to the details of its implementation in MATLAB and, probably, not enough effort invested in tuning its parameters. In the end all the results in all cases reported were achieved by the Pattern Search. The different modular modelling techniques were evaluated based on the different performance criteria presented in Chapter 2.

Table 5.4: *ANN model structures and training parameter for 1 step forecast. (F = Forecast horizon)*

Basin name	B1 (Bagmati)	B2 (Sieve)	B3 (Brue)
Network structure (nodes)	$5-4-1\ F = 1d$	$6-5-1\ F = 1h$	$6-14-1\ F = 1h$
		$4-5-1\ F = 3h$	$8-8-1\ F = 2h$
		$2-5-1\ F = 6h$	$5-27-1\ F = 3h$
			$4-24-1\ F = 6h$

5.6   Results and discussion

The performance measures of the three modular models schemes used are compared in Tables 5.5 and 5.6; all model forecasts are for one time step ahead. This analysis is important since the conceptual models also make a simulation for one step ahead.

The calculated RMSE in the three case studies show variable performance (Table 3). Modular models are in general better than global models but show variable performance. In the case of the more complex basin B1, with the largest area and largest forecast horizon, modular models improve on global models in relative terms more than for other basins. The mountainous region and the size make it a highly nonlinear system, and this implies that there is probably a large influence of baseflow components in the forecast streamflow and, consequently, higher importance of modelling it by a specialized model. The highest performance is shown by the MM3 model (RMSE lower than that of the global ANN by almost 24%). Although, MM2 performance measures does not seem to show a significant improvement with respect to all the error measures, it is possible to see that its error is consistently lower than that of the GM. With respect to the physical based models all the models clearly have better performance that the conceptual hydrological models for one time step forecast.

Table 5.5: *Performance in verification of the different modular models and global models for each basin (1 time step forecast).*

	RMSE					CoE			
	GM	MM1	MM2	MM3	Naïve	GM	MM1	MM2	MM3
B1	134.41	113.79	116.74	**98.55**	153.54	0.66	0.77	0.74	0.8
B2	3.71	5.84	3.41	**3.07**	6.73	0.9942	0.9855	0.9957	0.996
B3	0.11	0.15	0.11	0.11	0.25	0.9988	0.9959	0.9975	0.9976

Basins B2 and B3 are small in size with a relatively fast response. In general, it can be said that these catchments were modelled with high accuracy by both global and modular models. High accuracy makes it difficult to compare the models using CoE which is very close to 1 (Table 5.5). Nevertheless, it is clear that the use of hydrological knowledge in flow separation gives good results for these basins as well. The MM3 models show the largest reduction in RMSE

Table 5.6: *PERS indexf of different modular models and global models for each basin (1 time step forecast).*

	PERS			
	GM	MM1	MM2	MM3
B1	0.23	**0.47**	**0.41**	0.51
B2	0.69	0.24	0.78	0.79
B3	0.78	0.64	0.78	0.79

compared with the global model in B2; however, there is no clear conclusion for B3.

The results show that the error of the modular and the global models are less than that of the naïve model. The naïve model is the simplest solution and could be interpreted as a measure of the simplest form of linearity in the time series. All other models include the precipitation as an input variable (which the naïve model does not), so it is not surprising that they have better performance. It is also worth noting that the relatedness (measured by correlation) between the precipitation and the future values of discharge is variable and our experiments (not presented here) show that it depends on the different seasons (since under different flow conditions soil moisture and the time lags are also different). This prompts the idea of using different modular model structures for different seasons, and combining them in an overall model –which can be undertake in future research.

In terms of the coefficient of persistence (PERS, Table 5.6), the results are consistent with the RMSE and CoE. In terms of PERS, MM3 outperforms all other models in all three case studies. The PERS index for the MM3 is near or above 0.5, showing a significant increase in performance over the naïve predictor.

Using more than one error metric in the analysis makes it possible to better evaluate the performance of models for various hydrological regimes. In this study RE (Equation 2.10) is used to identify the percentage of samples belonging to one of the three groups: "low relative error" with RE less than 15%, "medium relative error" with RE between 15 and 35%, and "high relative error" with RE higher than 35%. The ranges were determined after experiments with the two trial models. The low error value is expected to cover possible measurement errors that could be around 20% (Beven, 2003). The percentages of samples in the three different relative error (RE) groups are shown in Figure 5.5. RMSE of MM2 for B1 is less than that of the GM, but at the same time there are fewer samples with low RE than for the GM. This seems contradictory, but the RMSE squares the absolute error so the high flow samples with low RE may have high absolute error, which, being squared, will contribute considerably to the total RMSE. At the same time, since their RE is low, they would be attributed to the "low relative error" group which would contain a large number of such examples. This is what happened with the samples pre-

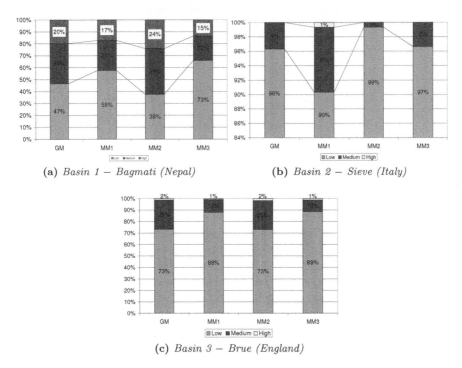

(a) *Basin 1 − Bagmati (Nepal)*   **(b)** *Basin 2 − Sieve (Italy)*

(c) *Basin 3 − Brue (England)*

Figure 5.5: *Relative errors classifid in groups of low (< 15%), medium (15% < RE < 35%) and high (> 35%)*

dicted by GM for which the "low RE" group appeared to be larger than for MM2.

For basin B2 there is an increase in the low relative error percentages for models MM2 and MM3 (Figure 5.5b). This is consistent with the NRMSE measure. In this case the MM2 model shows a more precise and accurate result having 99% of the sample with a low RE.

Figure 5.7 shows a fragment of a hydrograph generated by MM3 for B2 using test data. This model has been optimized, so that BFI_{max} is 0.25 and the recession coefficient a is 0.96. Indeed, for the fast response basins, values of this order are expected: the BFI should be small, and the recession coefficient should be high due to the relatively high slope in the recessions.

In contrast to B2, the B1 catchment in Nepal, which is large, has a considerable groundwater storage. This basin has a BFI_{max} of 0.95 and coefficient a of 0.23. Note that BFI_{max} as defined by Ekhardt (2005) (Equation 3.5) is the maximum value of the baseflow index (BFI) which is defined as the total volume of baseflow divided by the total volume of runoff for a period of time (Wahl and Wahl, 1995). A value of 0.23 therefore does not mean that the volume of baseflow is 23% of the total volume.

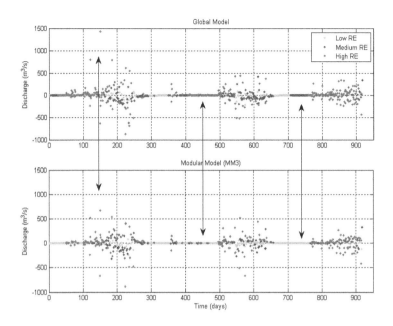

Figure 5.6: *Difference between measured and simulated, with highlighted relative errors for the global model (GM) and the modular model (MM) of the Bagmati river basin (B1)*

In general, the modular model MM3 outperforms the global model; this can also be illustrated graphically as in Figure 5.7a and 5.7b with a typical fragment of the hydrograph. Figure 5.7a shows that the baseflow for this catchment does not have a high contribution. This may actually explain why the accuracy of the modular model (where baseflow is modelled separately in this case) is not much higher than that of the global one. Another reason, of course, is that the GM is already very accurate.

An interesting question to ask is why the MM3 algorithm results in a better performance than that of MM1 and MM2. One may conclude that this can be attributed to the fact that Ekhardts filter is a better device to identify the baseflow, so the MM3 model is therefore better than the other models. However, the better performance of MM3 may also be a result of other factors, so that the further analysis is required. A more general question is whether the flow components identified by the separation algorithms really do correspond to different sub-processes (which we tried to model separately), or do these algorithms produce "baseflow" while not necessarily representing a clearly identifiable sub-process? One may argue, however, that accurate separation of sub-processes corresponding to base and excess flow (which are currently defined in a quite approximate fashion, and differently by different authors) is simply not possi-

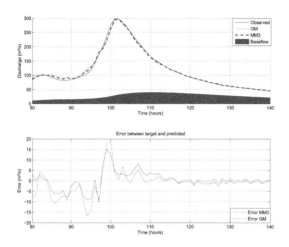

(a) *MM3 performance in basin B2*

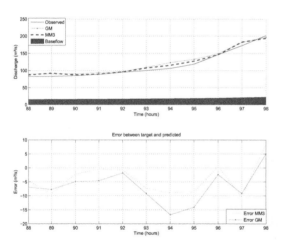

(b) *Performance comparison of MM3 and GM models in basin B2 (fragment).*

Figure 5.7: *Hydrograph section of the Sieve basin (B2)*

ble in principle. However, answering these interesting questions is beyond the scope of this work.

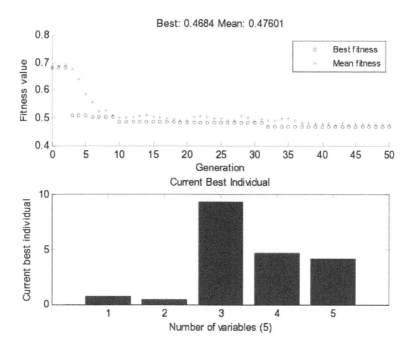

Figure 5.8: *GA optimization results, MM3 model, Bagmati case study (Variables: 1=BFImax, 2=a, 3=Bfo, 4=N1, 5=N2)*

Optimization

MM3 model optimization was performed using GA which was set up using the parameters shown in Table 5.7a. The convergence of the GA can be seen on Figure 5.8. One can see that GA converged relative fast to a quite low model error but was running for quite many iterations since at each of them a certain improvement was still observed. The total optimization time for the Bagmati case was around 3 hours (Pentium 4 processor running at 3.2 GHz was used).

The Generalized Patter Search (GPS) parameters are presented in Table 5.7b. GPS-based optimization was much faster than that by GA and for Bagmati case took around 10 minutes More accurate comparisons between these two and other direct search algorithms in the considered class of optimization problems are yet to be performed.

Models performance for different forecast horizons

The comparative analysis of models with respect to their performance for different forecast horizons was done for basin B2 (Sieve, Italy) and B3 (Brue, England). Figure 5.9 indicates that difference in performance between all mo-

Table 5.7

(a) *Genetic algorithm Optimization parameters*

Elite Count	4
Crossover Fraction	0.3
Population size	20
Generations	50
Initial population	Random
Selection Function	Stochastic Uniform
Cross over function	Scattered
Mutation Function	Gaussian
Mutation Function	Adapt Feasible

(b) *Generalized pattern search parameters*

Tolerance Mesh:	0.000001
Tolerance Function:	0.000001
Tolerance Bind:	0.001
Maximum Iterations:	500
Maximum number of function Evaluations:	10000
Mesh Contraction:	0.5
Mesh Expansion:	2
Initial Mesh Size:	1
Initial Penalty:	10
Penalty Factor:	100
Poll Method:	GPS positive basis 2N
Polling Order:	Consecutive

dular models and GM's accuracy becomes higher with the increase in model complexity. This points to the potential of using modular modelling in situations when forecasting (and modelling as a whole) becomes a difficult task, for example in critical flooding situations when higher forecasting horizons are desirable. It is also important to notice that the mentioned system complexity does not relate to the seasonality in the time series, an aspect that is still to be analyzed and addressed.

Two important remarks can be made in comparison with the performance of previous models like M5 prime and the conceptual models mentioned as benchmarks. It would appear that the accuracy of all the data-driven models is better than the conceptual models mentioned. On the other hand, for extended lead times, the hydrological conceptual models with an accurate input may have a better performance than the data-driven ones. It is worth stressing again that the modular models presented here perform better than the benchmark global models mentioned at the beginning of this section (Corzo and Solomatine, 2006a,b, 2005; Corzo et al., 2007; Solomatine and Dulal, 2003).

5.7 Conclusions

In this study the modular modelling approach to build hydrological forecasting MLP ANN models was used. Accordingly, instead of training a single (global) data-driven model on the whole data set, the training set is partitioned into several subsets, and a number of local models, each responsible for a region of the input space, are built. Three different partitioning schemes were employed: based on the clustering, on a traditional baseflow separation method (which was however updated to allow for algorithmic implementation), and on using

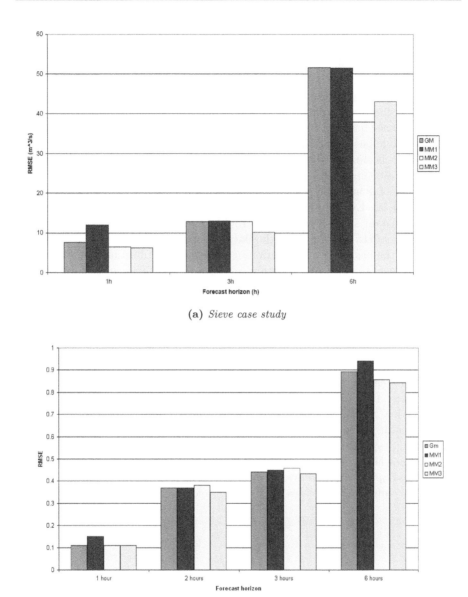

(a) *Sieve case study*

(b) *Brue case study*

Figure 5.9: *Comparison of the models performance for different forecast horizons*

the optimal hydrological process filter (which was optimized by the GA and
Generalized Pattern Search, with the higher performance of the latter).

The use of hydrological (domain) knowledge incorporated in the algorithms for separation of the base flow proved to be effective. Since such algorithms cannot be directly used in operation (they require the future values of flow), they have to be replicated by surrogate classifier models, and it was shown that this approach can be successfully implemented. Several classifiers used in this role were compared in accuracy but the difference appeared to be marginal (linear classifier was, however, the best). Partitioning the data by clustering in the input space has lead to less accurate models, if compared to those based on the knowledge-based partitioning (flow separation). However, clustering-based partitioning is simple, is not sensitive to the algorithm's parameters and can be used as a complementary tool.

The use of domain knowledge in the modelling framework presented proved to be beneficial. Even the traditional semi-empirical flow separation algorithms, such as constant slope algorithm, can add to the accuracy of data-driven hydrological models. Partitioning the data by clustering gave good results only in some of the basins. Such partitioning is simple, but does not directly relate to hydrological regimes and is highly sensitive to the distance measure used in clustering.

There are several research issues that are to be addressed (and which are already being addressed in the ongoing research): the proper "soft" combination of the modelling modules especially at the transition point from one regime to another; complementing hydrological knowledge by the routines for automatic identification of regimes (for example, using the apparatus of hidden Markov models); modularization of data-driven hydrological models (following, for example, an approach outlined by Solomatine 2006; Solomatine and Corzo 2006); and combining data-driven and physically-based models.

In general, it can be concluded that the modular approach for hydrologic forecasting, especially the one involving the domain knowledge, is useful. Use of such knowledge in partitioning the data and building local specialized ANN models and optimizing the overall model structure ensure accurate representation of the sub-processes constituting a complex natural phenomenon.

SPATIAL-BASED HYBRID MODULAR MODELLING, WITH APPLICATION TO THE MEUSE RIVER BASIN

The previous chapters of this thesis have shown that using the concept of hybrid and modular modelling leads to more accurate models. The concepts presented in Chapter 5 showed that, in a lumped model, the temporal and process modularization have an added value in terms of accuracy and richness of the embedded information about the forecasted situation.

In Chapters 2 and 3 the possibility of building models based on spatial modularity has been mentioned, and classified as class D2P (data-driven to process-based). This chapter explores such possibility in detail, on an example of a semi-distributed hydrological model of a meso-scale catchment (Meuse river basin).

6.1   Introduction

It is common practice to use semi-distributed conceptual models in operational forecasting for meso-scale catchments. These models are based on the principle of mass conservation and simplified forms of energy conservation. Conceptual models, however, may not represent all sub-basins with the same accuracy. Inaccurate precipitation data and the need for its averaging for the lumped models may seriously influence the accuracy of modelling. Due to the limited representation of the full rainfall-runoff process, the complexity of the model integration and the identification of the lumped parameters, the proponents of fully distributed detailed models argue that there are many situations when the accuracy of conceptual models is not sufficient. However, the simplicity of these models and the high processing speed is an advantage for real time operational systems and often makes such models the first choice.

Precipitation forecasts are normally available for low resolution grids which are close to the size of the modelled sub-basins. It has been shown that there are situations when such models are more accurate than the fully spatially distributed physically based and energy based models (Linde et al., 2007; Seibert, 1997).

Diermansen (2001) presented an analysis of spatial heterogeneity in the runoff response of large and small river basins, and an increase of error is observed with the increase of the level of detail in the physically based model. An alternative to fully-distributed models is the class of intermediate models, the so-called semi-distributed conceptual models, as the most appropriate modelling approach for meso-scale operational forecasting. In this research the IHMS-HBV model (Lindström et al., 1997) belonging to this class is used (http://www.smhi.se). In this chapter it will be called simply HBV, and will refer to the initial hydrological model formulation used as a hydrological prototype module in the flood early warning system for the rivers Rhine and Meuse.

Ashagrie et al. (2006) presented a long term analysis for the effects of climate change and land use change on the Meuse river basin using the HBV model. This analysis showed that the agreement between the observed and measured discharge is generally good, in particular flood volumes and the highest peak are simulated well. However, there are some problems with the medium flow (shape and peak values), and a systematic deviation for certain observed periods (i.e. 1930-1960) was also observed. de Wit et al. (2007b) explored the impact of climate change on low-flows. They found high accuracy for the monthly average discharge and for the highest (January) and lowest discharge (August), but there was an overestimation and underestimation observed in spring and autumn, respectively. Many performance calibration techniques with different types of models have been used for the Meuse. Booij (2005) presented the manual calibration and validation of the HBV based on expert tuning of model parameters. The problems mentioned above still remain unresolved and under investigation by a number of authors.

As has been already discussed, an alternative approach to flow forecasting is using data-driven models (DDM). Traditionally, modellers build a general model that covers all the processes of the natural phenomenon studied (overall model). In many applications of data-driven models, the hydrological knowledge is "supplied" to the model via a proper analysis of the input/output structure and the choice of the adequate input variables. These models are less sensitive to precipitation and temperature information in hydrological systems where high autocorrelation is found in streamflows. Therefore, in operational systems where missing data is an issue, such DDMs can replace the local sub-basin models. Additionally, a complex distributed water system requires local model evaluations and integration of models. So an alternative is to build an overall DDM for the whole basin, and for semi-distributed hydrological modelling a combination of hydrological process-based and data-driven models can be used. Additionally, the routing model integrating sub-basin models can be replaced by a DDM as well. These two approaches are explored in this chapter.

While our study of using modular models of the Meuse basin has been in progress, Chen and Adams (2006) presented the description of sub-basin models and their routing through the use of an ANN-based integration model. The basin area was around 8500 km^2, with a division into three sub-basins based mainly on the river network system. The calibration process included two stages: first, the whole catchment was considered (no sub-basin discharge information was available), and, second, with the use of output discharges from the basins to the outlet. This approach is similar to the one tested in the present chapter, but we considered a more complex basin, compared the model with the ANN routing integrator with a full basin hybrid model involving ANN submodels, and performed additional analysis of the variations of the models seasonal performance.

The objectives of this chapter are: (i) to analyse the performance of DDMs in their role as sub-basin replacements, in terms of local and overall flow simulation errors; (ii) to explore different data-driven methods as alternative methods for the integration or replacement of sub-basins; iii) draw conclusions about the applicability of the hybrid process-based and data-driven models in operational flow forecasting.

The outline of this chapter can be seen in Figure 6.1. Section 2 describes the hydrological characteristics of the hydrological semi-distribtued model of the Meuse river basin and its validation information. Section 3 describes the methodology of the 2 schemes. Sections 4 and 5 cover the application results of scheme 1 and 2 respectively. Section 6 discusses the results of both schemes. Section 7 presents the conclusion of the different experiment results.

6.2   HBV-M model for Meuse river basin

The conceptual hydrological model HBV was developed in the early 1970s (Bergström and Forsman, 1973) and its versions have been applied to many catchments around the world (Lindström et al., 1997). HBV describes the most important runoff generating processes with simple and robust procedures. In the snow routine, snow accumulation and melt are determined using a degree temperature-index method. The soil routine divides the forcing by rainfall and meltwater, into runoff generation and soil storage for later evaporation. The runoff generation routine consists of one upper non-linear reservoir representing fast and intermediate runoff components, and one lower linear reservoir representing base flow. Runoff routing processes are simulated using a simplified Muskingum approach and/or a triangular equilateral transfer function (Ponce et al., 1996).

HBV is a semi-distributed model and the river basin can be subdivided into sub-basins (HBV-S). This model simulates the rainfall-runoff processes for each sub-basin separately with a daily or hourly time step. Each sub-basin is divided into homogenous elevations which are then divided into vegetation zones. Further details about the HBV model can be found in Lindström et al. (1997)

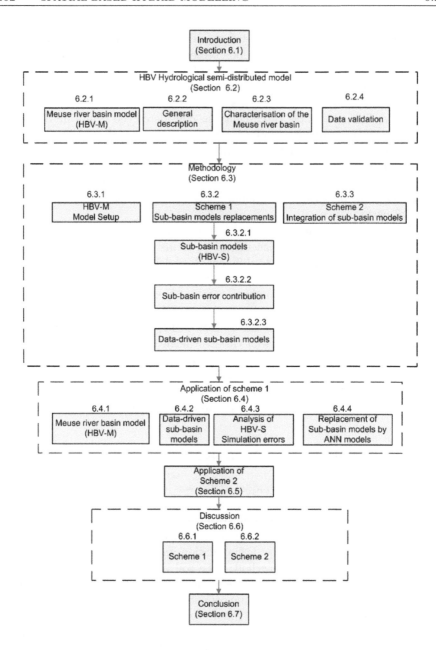

Figure 6.1: *Outline of Chapter 6*

and Fogelberg et al. (2004).

The HBV-S sub-basin models are linked by a Muskingum-Cunge equation.

This routing equation is conventionally applied to river reaches where the distance between the outlets of the basins is significant. The Equation 6.1 was used in this study.

$$Q_{n+1} = C_0 I_{n+1} + C_1 I_n + C_2 Q_n \tag{6.1}$$

$$C_0 = \left(\frac{K_x - 0.5\Delta t}{K - K_x + 0.5\Delta t} \right) \tag{6.2}$$

$$C_1 = \left(\frac{K_x + 0.5\Delta t}{K - K_x + 0.5\Delta t} \right) \tag{6.3}$$

$$C_2 = \left(\frac{K - K_x - 0.5\Delta t}{K - K_x + 0.5\Delta t} \right) \tag{6.4}$$

where, K is a storage factor with units of time, and Δt is the time interval considered in the simulation. The value of x represents the position on the river channel in meters. I_n and I_{n+1} are the input to the channel at the beginning and the end of the period Δt, respectively.

Diermansen (2001) presented an analysis of spatial heterogeneity in the runoff response of large and small river basins, and an increase of error is observed with a high increase of the level of spatial details in the model. An alternative to fully-distributed models is the class of intermediate models, the so-called semi-distributed conceptual models, as the most appropriate modelling approach for meso-scale operational forecasting. In this research the IHMS-HBV model (Lindström et al., 1997) belonging to this class is used (http://www.smhi.se). In this study it will be called simply HBV, and will refer to the initial hydrological model formulation used as a hydrological prototype module in the flood early warning system for the rivers Rhine and Meuse.

Ashagrie et al. (2006) presented a long term analysis for the effects of climate change and land use change on the Meuse river basin using the HBV model. This analysis showed that the agreement between the observed and measured discharge is generally good, in particular flood volumes and the highest peak are simulated well. However, there are some problems with the medium flow (shape and peak values), and a systematic deviation for certain observed periods (i.e. 1930-1960) was also observed. de Wit et al. (2007b) explored the impact of climate change on low-flows. They found high accuracy for the monthly average discharge and for the highest (January) and lowest discharge (August), but there was an overestimation and underestimation observed in spring and autumn, respectively. Many performance calibration techniques with different types of models have been used for the Meuse. Booij (2005) presented the manual calibration and validation of the HBV based on expert tuning of model parameters. The problems mentioned above still remain unresolved and under investigation by a number of authors.

6.2.1   Characterisation of the Meuse River basin

The Meuse River originates in France, flows through Belgium and The Netherlands, and finally drains into the North Sea (Figure 6.2). The river basin has
an area of about 33,000 km^2 and covers parts of France, Luxembourg, Belgium, Germany and The Netherlands. The length of the river from its source
in France to the North Sea at the Hollands Diep (an estuary of the Rhine
and Meuse rivers) is about 900 km. Major tributaries of the Meuse are the
Chiers, Semois, Lesse, Sambre, Ourthe, Amblve, Vesdre and Roer. The hydrological model of the Meuse basin upstream of Borgharen is subdivided into
15 sub-basins, covering an area of 21.000 km^2 (Figure 6.2). For more detailed
information about catchment geological and hydrological properties see Berger
(1992), de Wit (2009) and de Wit et al. (2007a).

In general terms the land use in the basin is made up of 34% arable land,
20% pasture, 35% forest and 9% built up areas (source: CORINE). Tu et al.
(2005) found the coverage of forest and agricultural land relatively stable over
the last ten years, but the forest type and management practices have changed
significantly. In addition to this it seems that intensification and upscaling of
agricultural practices and urbanization are the most important land changes
in the last century.

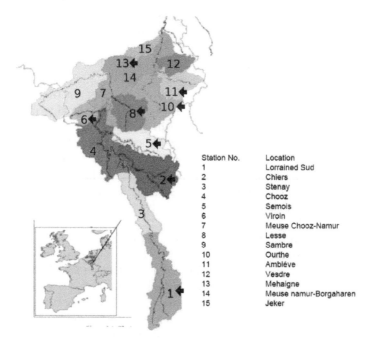

Figure 6.2: *The Meuse river basin and the sub-basins upstream of Borgharen*

As far as the hydrologic properties are concerned the Meuse can roughly be

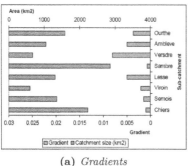

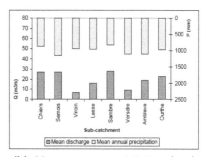

(a) *Gradients*   (b) *Mean average precipitation (mm)*

Figure 6.3: *Meuse basin hydrological properties*

split into three parts (Berger, 1992):

1. The upper reaches (Meuse Lorraine), from the Meuse source to the mouth of the Chiers. Here the catchment is lengthy and narrow, the gradient is small and the major bed is wide. Because of that the discharge up to the mouth of the Chiers has a comparatively calm course.

2. The central reaches of the Meuse (Meuse Ardennaise), leading from the Chiers to the Dutch border near Eijsden. In that section the main tributaries are Viroin, Semois, Lesse, Sambre and Ourthe. Here the Meuse transects rocky stone, resulting in a narrow river and a steep slope. The poor permeability of the catchment and the steep slope of the Meuse and most of the tributaries contribute to a fast discharge of the precipitation. The contribution of the area to flood waves is great, the contribution to low flows is small.

3. The lower reaches of the Meuse, corresponding to the Dutch section of the river. The lower reaches themselves may again be split into the stretches from Eijsden to Maasbracht and from Maasbracht to the mouth. In the former part the slope is still relatively high. For the greater part the river has no weirs here. In the section the Meuse has no dikes. For those reasons the stretch above Maasbracht is occasionally reckoned as part of the Meuse Ardennaise, which in that case flows from Sedan to Maasbracht. It may be remarked that the stretch that forms the border with Belgium is called the Grensmaas (Border Meuse) in the Netherlands, and Gemeenschappelijke Maas (Common Meuse) in Flanders.

6.2.2   Data validation

The validation of the data sets presented in this chapter are based on the results obtained from different researches (Ashagrie et al., 2006; Booij, 2002; de Wit et al., 2007b; Leander and Buishand, 2007; van Deursen, 2004). The overall water balance error obtained in the validation was ± 5%. Ashagrie et al. (2006) concluded that the average correlation of the HBV predictions and measured data is around 0.9, and the Nash-Sutcliffe efficiency is 0.93.

Hereafter HBV-M (HBV-Meuse) refers to the instantiation of the HBV rainfall-runoff model for the whole of the Meuse basin. The calibration and validation data sets used in HBV modelling were constructed in such a way that the observed and simulated discharges in both data sets in terms of flow volumes, and the number of flood peaks and the overall shape of the hydrographs are similar. However, initially no specific low-flow indices are used neither for calibration nor validation. Therefore in this study the results of the hydrological simulation of the Meuse discharges done by de Wit et al. (2007b) are used. In their study, the model was specifically validated against low-flow indices derived for the period 1968 to 1998.

Complementary information on data validation can be found in the research done by de Wit et al. (2007a). Their work presents the complete and detailed description of the hydrological data used for the model development.

6.3   Methodology

In this study two hybrid modelling schemes were tested. In the first one, some HBV-S (sub-basin) models were replaced by data-driven model representations. The second scheme is based on the replacement of the Muskingum-Cunge flow routing model by an ANN model integrating the outputs of the sub-basin models.

6.3.1   HBV-M model setup

The HBV-M model simulates the rainfall-runoff processes for each sub-basin separately. The sub-basins are interconnected within the model schematization and HBV-M simulates the discharge at the outfall 6.4. The schematisation and parameter optimization is derived from the approach proposed by van Deursen (2004).

The Meuse basin model has been calibrated and validated using daily temperature (T) and precipitation (P) for 17 locations interpolated from measurement stations, the calculated potential evapotranspiration (E_{pot}) per subbasin, and the discharge (Q) at Borgharen. The interpolation of the different locations was performed using Kriging (Stein, 1999).

HBV-M has been run on a daily basis using daily temperature, precipitation, potential evapotranspiration and discharge data for the period 1968-1984 (calibration) and 1985-1998 (validation) by Booij (2002, 2005) and fine-tuned (with more detailed data) by van Deursen (2004).

The model results in this study have been evaluated against the observed discharge records using (a) the volume errors (mm/yr), (b) the coefficient of

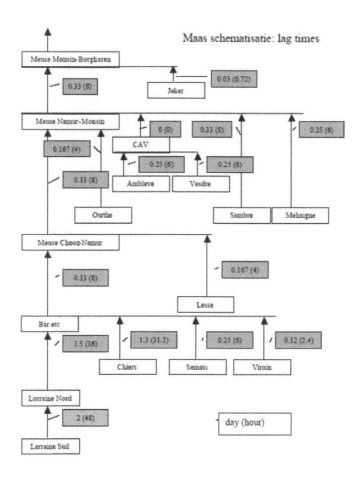

Figure 6.4: *Schematisation of the lag time considered between basins*

efficiency (CoE) for the gauging stations along the Meuse and the outlets of
sub-basins and the root mean squared error (RMSE, Eq. 2.1); (c) the nor-
malised RMSE (NRMSE, Eq. 2.3) (for comparing the sub-basin models with
considerably different flows).

6.3.2   Scheme 1: Sub-basin model replacement

HBV-S sub-basin models

The objective of the further analysis is to determine the average error contributions of the different sub-basin models (referred to as HBV-S) to the total
error of HBV-M, and hence to identify the candidate sub-basin models that
would need improvement or replacement. In this modelling exercise the river
basin behaviour during different seasons and flow regimes will be also taken

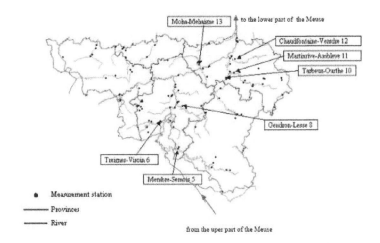

Figure 6.5: *Location of the river gauge stations available for the model replacement*

into account.

Sub-basin error contribution

The relative error contribution from a particular HBV-S (sub-basin) model is calculated as follows. First, the HBV-M model is run and its RMSE at the outlet is calculated. Then, according to a given replacement scenario a number of input measured discharges are fed into the HBV-M. These measured discharges were available only for some basins, and are the ones used for the different HBV-S model replacements scenarios. The HBV-M model is run once for each scenario. The resulting RMSE for each scenario is compared to the RMSE of the standard HBV-M. This gives the possibility of identifying the overall error variation due to the sub-basin model simulation. Such an error contribution is calculated for the different flow conditions (e.g., dry and wet seasons).

The replacement of the sub-basin models is performed in sequence: starts with the Lorraine Sud in the direction downstream towards Borgharen, then one more sub-basin model is replaced, then yet another one, until all selected sub-models are replaced (ending at Borgharen). It is important to stress that the independent replacements of sub-basins will not allow for seeing the accumulative error reduction, which is necessary to have an overall idea of the total error of accumulative areas. Two important assumption are made to be able to visualize the error contribution. First, is that the compensation of errors when adding the basin is minimal in comparison to the error of the basin contribution. The second assumption is based on the additive linear error propagation

Table 6.1: *Data available for the Meuse tributaries (catchment area until Borgharen)*

Sub-basin	Location of measurement tributary/river	% Area[a]	Available data
Subbas 1	St. Mihiel - Meuse	12.1	1969-2005
Subbas 2	Carignan - Chiers	10.5	1966-2005
Subbas 3	Stenay - Meuse	6.5	1982-2005
Subbas 4	Chooz - Meuse	10.7	1969-2005
Subbas 5	Membre Semois	5.9	1968-2005
Subbas 6	Treignes Viroin	2.5	1974-2005
Subbas 7	Maas Chooz Namur	5.4	-
Subbas 8	Gendron Lesse	6.2	1968-2005
Subbas 9	Sambre	13.1	-
Subbas 10	Tarbeux - Ourthe	7.6	1988-2005
Subbas 11	Martinrive - Ambleve	5	1974-2005
Subbas 12	Chaudfontaine - Versdre	3.3	1992-2005
Subbas 13	Moha - Mehaigne	1.7	1969-2000
Subbas 14	Maas Namur Borgharen	7.4	-
Subbas 15	Jeker	2.2	-

[a]Catchment area until Borgharen

along the river basin. Assuming non-linear error propagation may lead to complications of interpreting the contributions since there are temporal dynamics that affect the non-linearity.

Data-driven sub-basin models

After the error contribution of the HBV-S models are identified, data-driven models (DDM) can be built for each of the sub-basin models under consideration. Various data-driven techniques are compared to select the representative and accurate DDM.

As candidates for data-driven modelling, several statistical and computational intelligence techniques were tested: ANNs, linear autoregressive models and M5 model trees. Their performances were compared to that of the existing HBV-M model. Apart from that, an attempt was made to recalibrate a number of local HBV models; however, the overall performance obtained was lower than that after the calibration of HBV-M as a whole, and these experiments are not presented here. A detailed reference of the algorithms used can be found in Haykin (1999) and Witten and Frank (2000)

In the case study, before identifying the relative error contribution of various sub-basin models, several types of the DDMs were compared for the 8 of 15 sub-basins. This made it possible to judge if DDMs are useful as HBV-S replacements.

Each data-driven rainfall-runoff model for the sub-basins uses precipitation and measured discharge as inputs, and the response discharge of the basin is generated for the moment T time steps ahead. The general DDM forecast formulation can be represented as follows (it is the same formulation as introduced

Chapter 3):

$$Q_{t+T} = f(P_t, P_{t-1}, P_{t-2}...P_{t-L}, Q_t.....Q_{t-M})$$ (6.5)

where the optimal lags L for precipitation(P) and M for discharge (Q) are obtained through model optimization (these can be different for various forecast horizons T, in our case AMI and correlation results are used); f is the data-driven regression model, and T is the forecast horizon (e.g. 1 day). In this research several data-driven models are tested; including linear regression model (LR, (Kachroo and Liang, 1992)), artificial neural networks (ANN, (Dawson et al., 2005) and M5 model trees (MT, (Solomatine and Dulal, 2003)).

Neural network are all trained using the same random seed, with Levenberg-Marquardt algorithm. The learning rate was set to 0.1, one hidden layer with sigmoid function, and one linear transfer function in the output layer are common properties of the models. The ANN models have been optimized using a cross-validation set for determining the number of hidden nodes.

Building M5 model trees followed the procedure presented by Witten and Frank (2000). The size of the trees is controlled by fixing of the minimum number of instances in linear regression models at leaves (e.g. four).

6.3.3   Scheme 2: Integration of sub-basin models

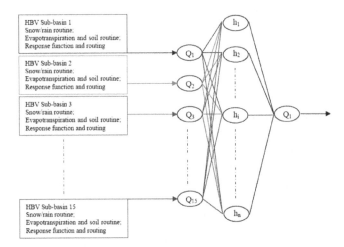

Figure 6.6: *Diagram of the ANN as replacement for the routing model, (Chen and Adams, 2006)*

Routing is a common way to integrate sub-basin models of a meso-scale catchment. However, river routing models include hydrodynamic conditions that require a large number of physical measurements. The accuracy is determined by the availability and the quality of these measurements and of the models. Since the cost of the measurements is high, often simplified routing equations are used. In HBV the sub-basin models use simple transfer functions that represent the routing process. However, to link HBV-S sub-basin models the Muskingham-Cunge equation (albeit simplified as well) is used. The routing equation is applied to river reaches where the distance between the outlets of the basins is significant.

The main idea of the Scheme 2 is the replacement of the traditional runoff routing by a more accurate non-linear function (data-driven model, Figure 6.6). In this chapter we have chosen for the multi-layer perceptron ANN (ANN-MLP) due to its widely known robustness and accuracy. The output discharges from the fifteen HBV-S sub-basin models are lagged and used as input to this model. The lags are determined using the correlation and average mutual information analysis involving different sub-basin flows and the final outflow at Borgharen. The ANN-MLP model has the following input-output structure:

$$Q_{Borgharen}^{t+T} = f(Q_{t-l_1^1}^1, Q_{t-l_2^1}^1, ..., Q_{t-l_M^1}^1, Q_{t-l_1^2}^2, ..., Q_{t-l_M^2}^2, ..., Q_{t-l_M^N}^N) \quad (6.6)$$

where the upper-index T represents forecast horizon, N is the total number of sub-basins, and l the lag at each sub-basin i. M is the number of lags taken per sub-basin i. All basins in the model are lagged with respect to the current flow at Borgharen.

6.4   Application of Scheme 1: data-driven models for sub-basin representation

6.4.1   Inputs selection and data preparation for DDMs

Each data set is split into a training set (70%; some data is used for cross validation as well) and a verification (30%) set. This procedure is performed in a way that ensures that the training data contains the maximum and minimum values of each variable to reduce the possible extrapolation problems. Additionally, the statistical similarity of each set was verified by comparing its probability density function.

The first step in developing data-driven models for the Meuse sub-basins was to identify the most appropriate inputs for predicting future discharges. Two approaches were used to select the appropriate input variables and their lags: correlation analysis and the average mutual information (AMI), as it was done, for example, by Solomatine and Dulal (2003) (Equation 4.2 and 4.3). A lag is defined as the number of time steps by which a time series is shifted relative to itself (when autocorrelated), or relative to the corresponding time values of another time series (when cross-correlated). The correlation coefficient and

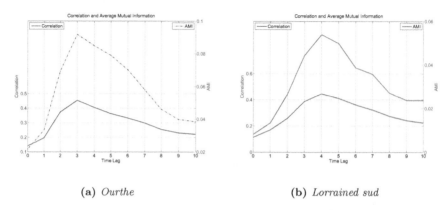

(a) *Ourthe*   (b) *Lorrained sud*

Figure 6.7: *Average mutual information for between lagged precipitation and discharge for sub-basins Ourthe (a) and Lorraine Sud (b)*

AMI were calculated for 10 lag values (Chapter 4). The variables compared were discharge, precipitation and evapotranspiration.

Based on a similar analysis to the one presented in the Figure 6.7, the following model structure was adopted for eight basins:

$$Q_t = f(P_t, P_{t-1}, P_{t-2}, P_{t-3}, Q_{t-1}) \tag{6.7}$$

The models were built for: Semois, Viroin, Lesse, Ourthe, Ambleve, Vesdre, Mehaigne, Chiers, Meuse Source; see their locations on Figure 6.2. The data used to build each sub-basin model (except Vesdre) covered the period from 1989 to 1995 for the training set and the period from 1996 to 1998 for testing. Due to the availability of data, for Versdre the data set used for training and testing covers the period from 1992 to 1996 and from 1997 to 1998, respectively. Stenay and Chooz (Sub-basin 3 and 4), have input from other three and one sub-basin flow (confluence sub-basins), and therefore not represented stricktly as catchment nor contemplated in this analysis. These two sub-basins are defined for the overall integrated HBV simulation and not for local model representation.

6.4.2   Data-driven sub-basin models

The performance of the HBV-S models was compared with that of several data-driven models (LR, M5P, ANN) (Figure 6.8); NRMSE was used as the error measure.

Both MLP and M5P data-driven models outperform the HBV-S models. Only for the Lesse, Ourthe, Ambleve, and Vesdre HBV-S model error is relatively low, but even then it is not comparable with that of the data-driven models. According to Berger (1992), Ourthe sub-basin together with Vesdre

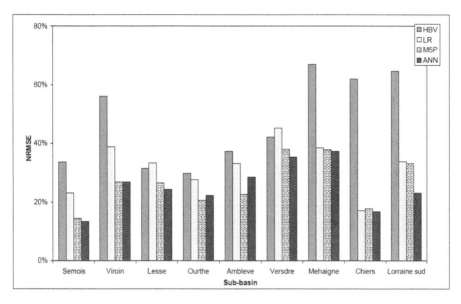

Figure 6.8: *Comparison of model performance for each sub-basin, expressed in NRMSE of streamflow (Calculated for verification period)*

and Ambleve are the most important tributaries for flood forecasting, relating area percentage and response time. HBV-M results for Semois, Viroin, and Mehaigne show high NRMSE. The error graphs show that the M5P and ANN models outperform the HBV model for all the considered sub-basins.

However, this does not mean that DDM is unconditionally superior to the conceptual modelling approach. The conceptual model aims to represent the processes of the modelled phenomena (albeit roughly), and the DDM is based on the analysis of historical data. Since the conceptual model only uses the forcing information (precipitation, temperature, etc), weather forecast information can be effectively used for the longer lead times. Other variables like measured discharges are incorporated in operational systems through the use of external post-processes like data assimilation.

The ANN-MLP model outperforms HBV in more cases than M5P does and therefore is selected for the replacement experiments. The results show that DDMs can serve as accurate replacement models for sub-basins. However, when more and more sub-basin models are replaced, there will be less and less hydrological knowledge (encapsulated in process models) left. In addition, for the extended forecast scenarios the weather information is highly important. Therefore an analysis of the overall performance of the model under different replacements is made below. Since there is a large number of possible scenarios of replacing various numbers of models, it is necessary to analyze the river basin behaviour and the relative quality of the individual HBV-S sub-basin models with respect to the overall basin measurements given by the discharge at Borgharen.

6.4.3   Analysis of HBV-S simulation errors

The changes in the overall model performance ($RMSE$) on the verification data set as a result of various replacements with measured discharge data are shown in Figure 6.9 and Table 6.2.

Table 6.2: *RMSE error contribution to the HBV overall simulation*

Sub-basin	Relative RMSE reduction (HBV-M)	Area (km^2)	Area (%)	Observed-Simulated (%Volume Difference)
Mehaigne	0.87	346	1.65	1.04
Ambleve	1.44	1 050	5.00	1.72
Ourthe	1.89	1 597	7.60	2.26
Lesse	2.36	1 311	6.24	2.81
Viroin	1.08	526	2.50	1.29
Semois	1.35	1 235	5.88	1.61
Chiers	3.79	2 207	10.51	4.53
Lorraine Sud	3.23	2 540	12.10	3.86
Others	67.82	10 188	48.51	80.89
Total HBV error	83.84	21 000	100	100

The replacement order can be followed by reading Figure 6.9 from top to bottom. From the total RMSE of 83.84, Chiers has the largest relative error contribution of 4.53% (10.5% of the total area), followed by Lorraine Sud and Lesse sub-basins with an error contribution of 3.86% (12.10% of the total area) and 2.81% respectively. Chiers is the second largest sub-basin of the Meuse and it is known that it commonly influences floods generated by its slow response, Lorraine Sud is also a slow responding basin. Vesdre, Ambleve, Viroin and Ourthe basins closer to the outlet are the most accurate in the HBV-M model and are the ones directly responsible for floods.

Hydrological data is available for 52% of the basin area; only 20% of the total errors seem to be attributed to this area. The rest of the error contribution can be associated with the other variables in the system, the modelling capacity of the HBV, as well as the different uncertainties in modelling of the basin. It would also be interesting to identify the error contribution of the Sambre, the largest sub-basin, but this was not carried out due to data unavailability. In Figure 6.9, the RMSE contributions obtained by each sub-basin replacement are associated with the measured discharge values.

Since it is well known that seasonality influences this river basin, the error contributions of the HBV-S models in summer (May October) and winter (November April) seasons are calculated in terms of the percentage of error with respect to the total HBV-M error; see Figure 6.9. The results in Figure 6.9 show that there is a homogeneous error contribution from Chiers in both seasons. The model for Lorraine Sud basin has a higher error contribution for summer and a small overall contribution in the winter. Clearly the calibration of the model is well suited for summer conditions where the slow response of the catchment is important for the average discharge in these periods. This is congruent with the size $(2540\ km^2)$, which represents approximately 10% of the considered area.

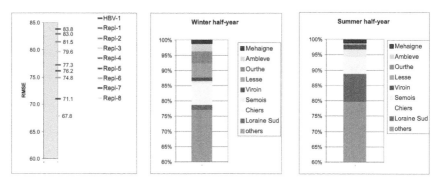

Figure 6.9: *Reduction in RMSE of the HBV-M due to accumulative replacements of the sub-basin models and seasonal performance*

In terms of flood forecasting at Borgharen the most sensitive basins for the HBV-M model distribution are Ourthe, Vesdre and Ambleve. The analysis shows that the Ourthe and Ambleve stream flows do not influence the model in the summer period, but together make a significant contribution to the error generated in the winter season. The contributions of the Mehaigne and Viroin sub-basins do not depend on the season: they have a small and similar error percentage for both seasons.

6.4.4 Replacements of sub-basin models by ANNs

There are numerous replacement scenarios and these should be identified based not only on the previous error analysis, but also taking into account the river basin behaviour during the different seasons and the different flow regimes. The total number of possible replacement scenarios (combinations of the sub-basin models with the data availability) is too high and it is not feasible to analyze them all. The experiments to replace a sub-basin model were carried out using only 8 scenarios as shown in Table 6.3.

The scenarios reflect mainly the fact that sub-basins with slow and fast flow responses contribute to different components of the resulting streamflow (mainly low and high flows, respectively). Characterisation of the eight scenarios (R1R8) is as follows:

- R1: The sub-basin (Chiers) with the largest error contribution, and a slow runoff response.

Table 6.3: *Replacement scenarios and the effect of their implementation.*

Short Name	Replacement	PAR^a (%)	ADC^b (%)	RMSE Reduction (ANN-S)	RMSE Reduction ^cMD	$\frac{ANN-S}{MD}$ (%)
R1	Chiers	11	10	3.84	4.53	0.85
R2	Chiers, Semois, Viroin	19	22	6.17	7.42	0.83
R3	Ourthe and Ambleve	13	15	2.25	3.98	0.57
R4	Ourthe, Ambleve, Lesse	19	21	4.21	6.79	0.62
R5	Ourthe Ambleve, Semois	18	25	3.73	5.58	0.67
R6	Semois, Chiers, Lesse	24	28	7.52	8.39	0.9
R7	Ourthe, Ambleve, Semois, Lesse, Chiers	35	41	8.62	12.92	0.67
R8	Lorraine Sud, Chiers, Semois	28	28	9.47	9.99	0.95

[a]Percentage of area replaced of the total basin (PAR)

[b]Average discharge contribution in relation to the total average discharge (ADC). The total average discharge is calculated using the average annual discharge from 1970 to 2000 (280.1 m^3/s).

[c]Measured data(MD)

- R2: Three sub-basins which include the Meuse tributaries upstream of Chooz. These are the highest elevation areas with relatively low slope and slow response during flood situations.

- R3: The two fast responding sub-basins that have high contributions during floods (Berger 1992).

- R4: The same sub-basins as in R3, but together with the slow response Lesse sub-basin whose model has a high error in summer and a low error in winter.

- R5: The same sub-basins as in R3, but together with the slow responding Semois whose model has a high error in summer and low error in winter.

- R6: Combination of slow and fast responding sub-basins.

- R7: Combinations of slow and fast responding sub-basins, but with a larger area covering 35% of the basin.

- R8: Slow responding sub-basins with a large total area.

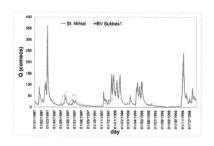

(a) *Sub-basin 1 (St. Mihiel - Loraine Sud)*

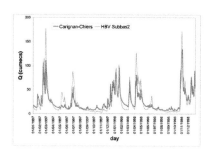

(b) *Sub-basin 2 (Chiers)*

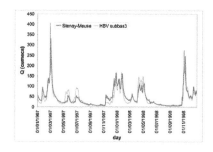

(c) *Sub-basin 3 (Stenay-Meuse)*

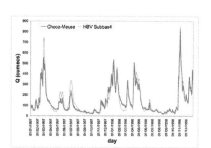

(d) *Sub-basin 4 (Chooz-Meuse)*

Figure 6.10: *Hydrograph of sub-basin models with important contribution to the overall model*

Table 6.3 presents the HBV-M model performance changes as a consequence of the different ANN-S replacement strategies. The following statements describe the interpretation of some of the results:

The effectiveness of the models replacements can be evaluated by analysing the changes in the overall HBV-M RMSE. The last column presents the percentages of the maximum reduction possible in case of implementing a particular replacement scenario.

Comparing sub-basins with similar area and similar discharge we can see where the replacement of models was more successful. For example. R1 and R3 have similar percentage of area (11 and 13 respectively), also similar average discharge contribution (10 and 15 respectively). However, the R1 (ANN-S) model gives a RMSE reduction (85%), which is higher than that for the scenarios R3, corresponding to larger areas and higher average discharge. This is an indicator that low flows play a significant role in the overall process, and also reflects the weakness of the HBV-S models currently used in simulating low flows. Other similar case can be seen when R7 replaced a bigger area (35%) than R8 (28%), however, the efficiency for the latter replacement is significantly higher (95%). In terms of discharge, R8 has a smaller average discharge and therefore less contribution. For the scenarios R6 and R8 results show a similar error reduction after the replacement. They have approximately the same average discharge percentage contribution to the basin and a similar area, however, their seasonal error contribution is different (Figure 6.9).

The influence of changing Ourthe and Ambleve for Lorraine Sud shows that most of the errors arise in the low flow modelling. The Lorraine Sud (location of the Meuse source) is the most distant basin with relatively mild slope, and therefore its contribution to flash flood (fast flow and runoff) is minimal. This is consistent with the results of de Wit et al. (2007b), who showed that the peak discharges of Vesdre and Ourthe basins are larger than those of Chooz. The results point to a partial explanation based on the differences in precipitation depths of the region and on the difference in hydro-geological conditions. On the other hand, the basins Ourthe and Ambleve (central part) are closer to the outlet and their individual performances are more sensitive for short time lags and fast phenomena.

The results of simulations for the verification period (last three years) are evaluated by calculating the RMSE and Coeff of efficiency (Figure 6.12). A typical section of the hydrograph is extracted in Figure 6.11. The shape of the hydrograph with ANN-S replacement is mainly driven by the overall hydrological model. It is possible to see that after the replacement R8 the flows (under 600 m^3/s) are closer to the observed discharge. For flows above 600 m^3/s the HBV-M is hardly affected due to the low influence of the replaced basins during the peak flow events (Figure 6.11). This shows that the replacement affects mainly the low flow simulation periods.

If one analyses only the reduction in the overall RMSE, then the replacement scenario R8 would result in the model that can be recommended to be used instead of the HBV-M.

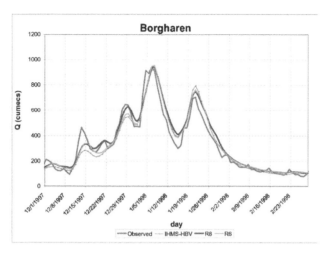

Figure 6.11: *Hydrograph replacements (R8) and (R6)*

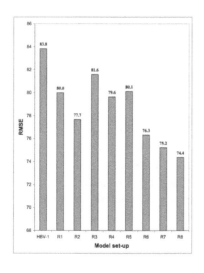

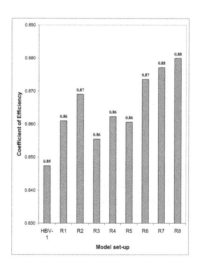

(a) *RMSE*   (b) *Coefficient of efficiency*

Figure 6.12: *Error reduction with different combinations of replacement (evaluated at Borgharen)*

6.5 Application of Scheme 2: integrating sub-basin models by ANN

To build a neural network model for routing, preprocessing and input variable identification is required. For this the AMI and cross-correlation analysis

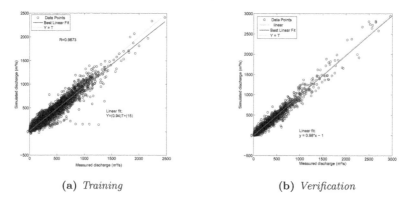

(a) *Training*   (b) *Verification*

Figure 6.13: *Scatter plots of target (measured) and ANN model for training and verification period*

were carried out to identify the relation (time lag) between the local sub-basins discharge calculated by the HBV model and the measured discharge at Borgharen. For most of the sub-basins the maximum value of AMI related to the observed discharge at Borgharen is a time lag of 1 day. Exceptions are sub-basins 9 (Sambre), 10 (Ourthe), 13 (Mehaigne), and 15 (Jeker) since the corresponding AMI is at maximum for lags less than 1 day. These results are in agreement with recent research (de Wit et al., 2007a), where it was found that the travel time between the measuring stations of the Sambre and Mehaigne, Ourthe and Jeker to Borgharen is less than half a day. More precise time lags can be obtained with hourly data. The average travel time of the flow between the Semois measuring station (sub-basin 5) and Borgharen is 1 day (Berger, 1992). The training data set used was from 01-01-1967 to 30-12-1988 (8035 samples) and for verification (3652) from 31-12-1988 to 31-12-1998.

The results of the model can be visualized by correlation graphs. High correlations are found between the observed and simulated discharge both for training and verification sets (Figure 6.13).

Figure 6.14 shows the observed and simulated discharges at Borgharen from December 2 1990 (record 700) to June 20,1991 . On average the integrated HBV-ANN model outperforms the original HBV-M model. The recession curve of the hydrograph is clearly closer to the measured curve and what was viewed as the systematic error in the recession curve of the HBV-M model is now corrected. An interesting phenomenon can be observed close to the measured peak: the measurement value goes up and down before it reaches its maximum value. This peak change in the hydrograph is reproduced by the ANN routing model with a relatively small underestimation.

For a 3 years error analysis the HBV-ANN gives RMSE of 58.66 m^3/s. An extended error analysis of nine years verification period shows that the RMSE

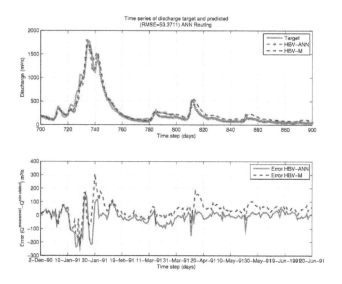

Figure 6.14: *Hydrograph of the original HBV-M and HBV-ANN integrated models*

for the HBV-M and HBV-ANN are 86 m^3/s and 55 m^3/s respectively which is
a 36% improvement (Table 6.4). The coefficient of efficiency is also improved
from 0.918 for the HBV-M to 0.967 for HBV-ANN model. For both winter and
summer seasons it is clear that the use of ANN for integrating the sub-basin
models improves the accuracy.

Table 6.4: *Comparison between the HBV-M model and the integrated HBV-ANN
model*

Model	Hydrological year (Nov-Oct)		Winter (Nov.-April)		Summer (May-October)	
	HBV-1	HBV-ANN	HBV-1	HBV-ANN	HBV-1	HBV-ANN
RMSE(m^3/s)	85.65	54,51	100.02	64.26	71.66	45.56
NRMSE	0.286	0,182	0.273	0.175	0.484	0.308

Integrating Scheme 1 and 2

The ANN-MLP routing model integrates the results of the sub-basin models
and generates the value of discharge at the outlet (Borgharen). By doing so,
the ANN routing is already correcting the regional behaviour of each sub-basin
model, so the ANN-MLP routing acts as an error corrector. Therefore, the use
of another sub-basin model (e.g. R8 scenario in Table 6.3), with different error
performance, as input of the ANN-routing model does not add new knowledge

into the model, but only increases the error. The replacement R8 into the
ANN-MLP (scheme) had almost the same performance as the original HBV
without any replacement (RMSE 82.91, see Fig. 6.15).

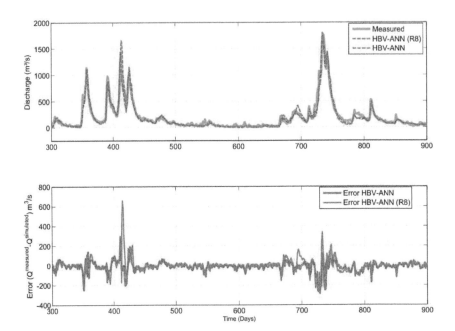

Figure 6.15: *Hydrograph comparison for the HBV-ANN with and without R8 replacement*

6.6 Discussion

In this section the results for each scheme are discussed and compared.

6.6.1 Scheme 1

The results show that replacing some of the conceptual sub-basin models with
data-driven models clearly improves the overall model performance. Doing so
the low flow errors related to some of the sub-basins can be reduced without any
deterioration in the high flow performance. The operational forecasting system
using process-based models requires variables like precipitation and temperature for each simulation forecast, however, with ANN models only previous
measured discharge is needed. Therefore, this approach may bring operational
advantages on the locations where weather forecast information may not be
available.

The choice of the best combination of HBV and ANN components depends on various factors and is, in fact, a multi-criteria problem. One may also think of rules (taking into account for example the season, data availability, location) that decision maker would use to select the final model.

The use of Scheme 1 may well be suited for simulation, but comparative tests with data-assimilation and data-driven approaches of the whole basin may be needed to determine whether the use of data-assimilation in operational system is more accurate or suitable than a simple ANN model of a basin. This analysis will be conducted in further studies.

Note that the extended forecast made by DDMs need the previous simulation discharges. Three important implications have to be mentioned here. First, the use of previous simulation discharge iteratively decreases the quality of the forecast. Second, if we assume that the measured information is a perfect forecast, the HBV average performance will not decrease for the higher forecast horizons. The third consequence is that the DDM is not representing the basin behaviour and instead is acting more as an autoregressive model.

The data-driven model, which tends to generate high weight values for input from previous discharges in its structure, underestimate the use of other variables that are poorly correlated with the output. In this sense data-driven models (DDM) can simulate the flow quite accurately (only on average, however, and not in the beginning of a high precipitation event) even without the use of the variables that really drive the phenomena (precipitation and temperature).

6.6.2 Scheme 2

Applying the integrating ANN model (Scheme 2) leads to a more accurate calculation of the overall river discharge, if compared to both Scheme 1 and to the simplified routing scheme employed in the HBV-M model. Our results in this experiment are in agreement with the work by Chen and Adams (2006) where an ANN model was used to integrate the three basin models (Xinanjiang model, Tank model, and Soil Moisture Accounting model).

The use of physical conceptual and data-driven models in operation should consider the dynamics of the basin. The dynamics of the Meuse basin has hardly changed during the last decade (Tu et al., 2005), so the combination of models seems to be reliable under relatively long periods of time (e.g. 3 to 5 years as the validation period of the models presented).

It should be noted that the experiments presented in this chapter are based on daily data and are aimed at improving the HBV-M hydrological model. In subsequent studies it is planned to explore the usefulness of the approaches above under a more detailed and complex framework (daily forecast with hourly data and precipitation forecast information). The challenge in extending these concepts to hourly-based models relates not only to the non-linearity and dynamics, but also to the influence of human interventions at weirs, sluices, canals, power plants, etc. These aspects are not included in the HBV-M model and are part of the motivation to use data-driven techniques, and, possibly, rule-based techniques allowing for multiple regimes of model operation.

6.7   Conclusions

One of the challenges of flow simulation is to increase the models accuracy. This chapter explored the use of data-driven models, e.g. artificial neural networks (ANN) to improve the flow simulation accuracy of a semi-distributed process based model. The IHMS-HBV model of the Meuse river basin is used in this research. Two schemes are tested in the modelling process: The first one explores replacement of sub-basin models by data-driven models. Error contributions to the overall flow simulation due to the use of ANNs for sub-basins are evaluated. The second scheme is based on the replacement of the Muskingum-Cunge routing model, which integrates multiple sub-basin models, by an ANN. The results showed the following: (1) after a step-wise spatial replacement of sub-basin conceptual models by ANNs it is possible to increase the accuracy of the overall basin model; (2) there are seasons where low and high flow conditions are better represented by ANNs; and (3) the improvement in terms of RMSE obtained by using of ANNs is greater than using sub-basin replacements. It can be concluded that the presented two schemes based on the analysis of seasonal and spatial weakness of the process based models can improve performance of the process based models in the context of operational flow forecasting.

As one of the following steps, it is planned to move from daily to hourly data, and from the one-step ahead forecasts to the models that forecast the flow several time steps (hours) ahead. In this case the set of inputs of DDMs may not include the previous values of flow (since they cannot be measured), and their performance may deteriorate if compared to the conceptual models that do not need the discharge as input and are fed with the precipitation forecast only. A possible answer could be in using the architecture of DDM that would use estimates of flow, or to use an ensemble of conceptual and data-driven models.

There is also a problem of limited and inaccurate data for most of the sub-basins, and this affects the performance of operational systems. A possible way to alleviate this, is to use autoregressive models which are not sensitive to the precipitation, temperature and evapotranspiration. As shown in appendix C, on an extended hourly forecast analysis of the Meuse river system (Meuse Delft-FEWS), it is possible to see how DDM models have better performance, on different regions of the semi-distributed system, for a significant high number of time steps. Yet another issue is the estimation of the models uncertainty associated with the inaccuracies in data and model structures. It is planned to explore all these issues and possibilities in further studies.

CHAPTER SEVEN

HYBRID PARALLEL AND SEQUENTIAL MODELS

Complementary models that work in parallel with the operational flow forecasting model and aim at reducing its error are becoming more and more popular during the last decade. One of such approaches uses a parallel architecture (model ensembles, i.e. several models running in parallel); the disadvantage is that it becomes difficult to speak about the "model state space". Another way of combining models is a sequential combination of models like it is done in a data-assimilation process that updates model states (or only its outputs). Contrasting these approaches as "parallel" and "sequential", and comparing their performance may be useful for selecting the model architecture for operational forecasting.

As mentioned earlier, the conceptual or process based models seem to be inaccurate in simulation for the fixed lead time. However, in hindcasting (using measured information in forecast time), such models would be very accurate for extended forecast horizons. In this Chapter the use of single global models, models operating in parallel (ensembles) and in sequence (leading to data assimilation by error correction) are applied to the Meuse rived basin for the multi-time step forecasting.

7.1 Introduction

In operational flow forecasting the information available from river gauges is typically assimilated into conceptual hydrological, but the assimilation procedure is outside of the model itself. Data assimilation can be done by updating internal model states or by a model that corrects the output errors. Both approaches are independent from the hydrological conceptualization of the process and do only base their corrections in optimal statistical updating strategies or error corrector models. The updating of model states is a common data assimilation approaches, however, in models using combinations of several conceptual models, the process becomes quite complex. Error corrector procedures have been explored in various publications (e.g., Broersen and Weerts 2005; Butts et al. 2002; Madsen and Skotner 2005) and nowadays are made part of most operational forecasting systems. These studies explored mainly the correction of the output based only on past errors. By using autoregressive (AR) linear and non-linear error corrector models they show the significant improvement of the hydrological conceptual model accuracy. At the same time most studies show that the difference in the performance between linear and non-linear error corrector techniques is not significant. Also, the relatively low effectiveness of such procedures for extended lead times was observed.

It is important to mention an important difference between hydrological conceptual (process-based) and most of the data-driven models used for flow forecasting. Typically conceptual hydrological models use forcing variables like precipitation and temperature to calculate runoff, and do not use past information about discharge. On the other hand, data-driven models often rely on the past measurements of discharge which are highly correlated with the output (see Chapter 4).

This chapter presents a comparative study of some of the mentioned types of models, with external variables (e.g. previous measurement and model results), in order to assess what are the advantages and disadvantages of various techniques for real time operational forecasting. The study is based on the daily forecasts performed by the existing HBV hydrological model of the Meuse river basin in the French and Belgium area (Chapter 6). Artificial neural networks (ANN), and linear regression models were used for output error correction. Two hybrid model composed of ANN and HBV models, linear and non-linear combinations, are presented. The results of the different models are evaluated with respect to the average daily discharge for one and multiple time steps ahead.

This chapter is divided in 5 sections. Section 2 presents the different methodologies used and setup of models applied to the Meuse river basin. Section 3 discusses forecasting considerations in the application of the different models. Chapter 4 presents the results and discussion of the different models performance. Conclusions are drawn in Section 5.

7.2   Metodology and models setup

The modelling approaches considered in this chapter are grouped in three: single forecasting models(ANN and HBV-M), error correctors (EC), and committee models (CM).

7.2.1   Meuse river basin data and HBV model

The HBV-M model setup described in section 6 is the basis for the different parallel and sequential models presented in this study. The location and description of the area used is described in Chapter 6. Temperature (at 15 sub-

basin centroids), precipitation (at 15 sub-centroids) and discharge from 1968
to 1998 were used for this chapter. Data was partitioned: 70% of continuous
samples were used for training and 30% for validation. The sub-basin centroids
are calculated as is mentioned by Ashagrie et al. (2006) for each one of the 15
sub-basins presented in Chapter 6. Since the characteristic response time in
this river basin is approximately one day, most forecasts will be analyzed first
for a one day forecast, and then for an extended forecast till 11 days. The
hydrograph of the river basin is shown in Figure 7.1. Equation 7.2 shows the
input used for the HBV-M models.

$$Q_{HBV_{i_t}} = f(P_{i_t}, Evap_{i_t}, T_{1_t}|S); \qquad (7.1)$$

$$Q_{HBV_{B_t}} = f(Q_{HBV_{1_t}}, Q_{HBV_{2_t}}, ..., Q_{HBV_{15_t}}); \qquad (7.2)$$

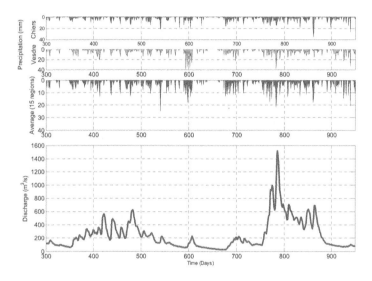

Figure 7.1: *Daily Meuse river basin hydrograph*

7.2.2   ANN model setup

To setup the ANN MLP model, a number of experiments considering different
correlated and with high average mutual information were tested; following
the procedure described in Chapter 4. The correlations found between the
precipitation data and discharge at the outlet of the river basin (Borgharen),
in the different basins showed that three closest sub-basins (Vesdre, Ambleve
and Ourthe) had the maximum lag at 2 days, and all other sub-basins 3 days.
This is consistent with the hydrological research done by Berger (1992). From
the correlations between temperature and discharge the maximum value found
was 0.15, which is not representative.

$$Q_{ANN(Q)_t} = f(Q_{m_{t-1}}, Q_{m_{t-2}}, Q_{m_{t-3}}) \qquad (7.3)$$

$$Q_{ANN(Q+P)_t} = f(Q_{m_{t-1}}, Q_{m_{t-2}}, Q_{m_{t-3}}, P_{1_{t-3}}, ..., P_{5_{t-3}}, P_{6_{t-2}}, ..., P_{15_{t-2}})$$
$$(7.4)$$

Where Q_{m_t} is the discharge measured at time t, for longer lead time steps an iterative approach is used; for past values of of time t ($t-1, t-2$ or $t-3$) the Q_{m_t} becomes $Q_{ANN(Q)_t}$ or $Q_{ANN(Q+P)_t}$.

In terms of average mutual information (AMI, Chapter 4), the analysis of precipitation lags versus discharge has the first peak (measure of maximum information) at the same time as the correlation analysis. Based on these results, the input variables for the three schemes were tested. These preliminary schemes were developed with only discharge (ANN (Q), 1), discharge and precipitation (ANN (Q+P), 2). The common input in the mentioned experiments was the use of discharge with 1, 2 and 3 days lag. The number of nodes was determined by exhaustive model optimization, varying the number of nodes from 1 to 30, and selecting the best performance on a cross-validation data set (10% of the training samples). The results confirm the high influence of the correlated input variables on the model performance. In the remaining part of this study the single ANN will be referred as ANN (Q) and ANN (Q+P) in Equation .

7.3   Data assimilation (error correction)

Data assimilation methods are quite developed, and were classified by the WMO (1992) based on the variables that are modified in the feedback. There are four options used to assimilate real time data (Figure 7.2).

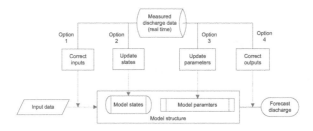

Figure 7.2: *Approaches for data-assimilation models in forecasting models; based on WMO (1992)*

1. *Updating of input variables*: the input variables are corrected based on the idea that noise in the measurement, interpolation, and other sources of errors are commonly present in the forecasting system.

2. *Updating states*: since the states are highly sensitive variables for the outcome of the model, this method uses the measured output of the model to update these variables. However, this approach explicitly modifies the assumption of the basin representation since the water balance on the conceptual (tank) model is modified

3. *Update of model parameters*: it is believed that the model could be dynamic and therefore requires correction on long term forecasts. This is not common since in task of operational forecasting the model is assumed to be stationary and hardly changes for a long period of time.

4. *Update of output discharge* : This method is used mostly since it does not require modifying the internal process in the hydrological model. The implementation is simple and most practitioners use simple linear regression models for that.

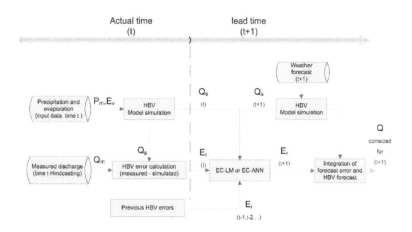

Figure 7.3: *Forecasting scheme using HBV process model with an output error corrector model*

Various authors report different results when the performances of different data-assimilation techniques are analyzed (Babovic and Fuhrman, 2002; Broersen and Weerts, 2005; Madsen et al., 2000). Madsen et al. (2000) compared global linear autoregressive models with artificial neural networks and genetic programming. Their results showed that the ANN error corrector was similar to the AR models. However, the best performance is commonly obtained by the Ensemble Kalman filter and an ANN, with a small difference between them. Based on this, our research will focus on the use of ANN as potentially effective nonl-linear error corrector model (EC-ANN, Figure 7.3). The error information is obtained for each time step (Equation 7.5) .To have a reference model, a single linear error corrector regression model is considered as well (EC-LM).

$$E_{m_t} = (Q_{HBV_t} - Q_{m_t}) \qquad (7.5)$$

$$EC_t = f(E_{m_{t-1}}, E_{m_{t-2}}, E_{m_{t-3}}) \qquad (7.6)$$

Where E_{m_t} is the actual error of the HBV model (time t), for longer lead time steps an iterative approach is used; for past values of of time t ($t-1, t-2$ or $t-3$) the E_{m_t} becomes $Q_{ANN(Q)_t}$ or $Q_{ANN(Q+P)_t}$.

To setup the error corrector model the correlation analysis is done. The autocorrelation of the HBV model errors is presented in Figure 7.4. There is autocorrelation of the error results till a lag of 2 or maximum 3 days, after this period, it may not provide important information for the autoregressive error corrector models. The equations used for the error corrector models are presented in Equations 7.1, 7.2, 7.5 and 7.6. Commonly the use of 7.6 is used in most error corrector models, however, in this thesis the exogenous(x) inputs are used as well.

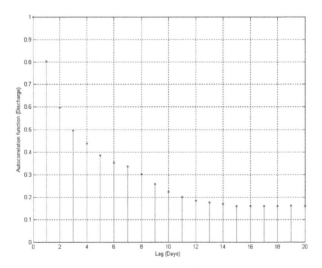

Figure 7.4: *Autocorrelation function of the HBV model of the Meuse river basin*

7.4   Committee and ensemble models

The ensemble of models is relatively new approach in rainfall runoff modelling; such approach is sometimes called ESP (ensemble of streamflow prediction). In this study we explore the combination of conceptual hydrological model (HBV) and ANN models. The elaboration of the models followed the analysis described for ANN models in Chapter 4. Two models are developed for the integration of HBV and ANN, a linear regression model and a non-linear ANN model.

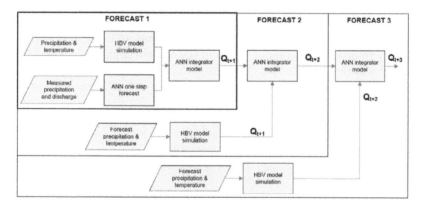

Figure 7.5: *Forecasting scheme of a committee model using ANN to integrate HBV and ANN*

A hybrid parallel scheme (committee machine) of HBV and ANN model (CM-ANN), for three time steps forecasts is shown in Figure 7.5. The model in the first time steps combines the ANN and HBV model results, but for time steps two and three, the last combined forecast, and last HBV forecast are used as input to the new ANN model combination.

$$CM_t = f(Q_{HBV_{B_t}}, Q_{ANN_t}) \tag{7.7}$$

The Equation 7.7 in continuous forecast uses the value of CM_t instead of Q_{ANN_t} to obtain CM_{t+1}. The function f in the experiments used here are ANN (CM-ANN) and LM (CM-LM), which are built up on the same number of training samples.

7.5   Forecasting scenario

In operational forecasting the hydrological model simulations are fed with precipitation forecasts to allow for extending the lead time. For the comparative analysis of the models, it is important to standardize the performance of the hydrological conceptual model in order to see the usefulness of the error correction schemes. To achieve this, the performance of the HBV model with the verification data set (unseen in calibration) is used as reference for any forecast. This can be interpreted as an assumption that precipitation forecast is perfect (measured in hindcast analysis).

The influence of precipitation in the HBV model of the Meuse river basin has been explored by Hidayat (2007). This work explored the influence of including random noise of different levels to the three of the sub-basin of the HBV Meuse model. This three basins have in average 20% of discharge contribution to Borgharen (outlet of the basin). With the noise, the HBV showed an increase of the overall RMSE error up to 4%. On the other hand, the ANN models with the same noise presented only 0.5% of increase on its RMSE error. The sensitivity to precipitation input errors in the conceptual model is considerable higher than in the ANN model. Therefore, the results obtained here, assuming the hydrological conceptual model performance as reference, is susceptible to change in a range of $\pm$ 20% difference.

From the DDMs perspective, to be able to extend the forecast into different lead times it is possible to iterate the model using its previous results or to create an independent model. In the other hand, they are set up based on mathematical correlations and expert knowledge of the forcing variables of the real world system. The ANN developed as single overall forecast models incorporate its own previous forecast discharge as input for each new forecast as mentioned above.

To determine the performance of the different modelling approaches it is required the use of different error metrics. In the assessment of the overall model ability the forecast error measures described in Chapter 2 are used.

7.6   Results and discussion

7.6.1   Single forecast results

The modelling approaches have been tested for one time step and on multiple time steps forecasts, from 1 to 11. The performance results of the basic reference models using one day ahead (approximately lag time of the basin) without data assimilation are presented in Table 7.1. It is clear that the process model has the lowest performance but it is important to highlight that it is the model that represents the runoff based only on the model actual internal state and the rainfall. The ANN trained with only Q seems to have better performance than the one which takes into account the precipitation; this results can be attributed to the high autocorrelation found in the discharge(decreasing from 0.97 at 1 day and 0.86 at 3 days).

Additionally, the exhaustive optimization procedure was used to determine the confidence interval of this ANN model RMSE results due to random initialization. For this the result of generated 30 ANN models were analyzed using

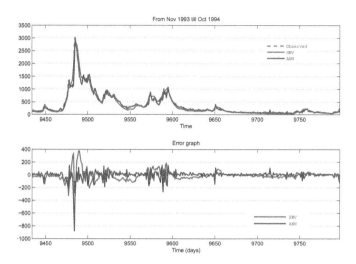

Figure 7.6: *Simulation results and error graph from HBV and ANN models for the year 1993*

Table 7.1: *Performance of the HBV and ANN models for the testing period 1 day ahead.*

	HBV	ANN(Q)	ANN(Q+P)
RMSE(m^3/s)	79.8064	55.5674	55.8426
CoE	0.9121	0.9364	0.9358
Cor	0.9573	0.968	0.9677
NRMSE	29.6555	25.2165	25.3414
MAE	50.1943	29.3466	30.1339
PERS	-0.3354	0.2042	0.1963
SSE	72117000	3501500	3536300

the Students t-test. The null hypothesis tested was the assumption of sampled data from a random normal distribution with mean 58.85. This hypothesis was analyzed against the alternative that the mean is not 58.85. The result of the test indicated a failure to reject the null hypothesis at the 5% significance level. The standard deviation of the process was an RMSE 3.95. This gave a confidence interval, of the RMSE obtained in the ANN model 58.05 ± 3.95. ANN models generated having RMSE above this intervals are rare. Based on this, using precipitation does not make a significant difference in performance compared to the average ANN model (Table 7.1); however, the error correction and committee models do. Note that this analysis concerns the mean error metrics (RMSE), and this could be misleading when particular (extreme) events are considered in which knowledge of precipitation could be a decisive factor for an accurate forecast.

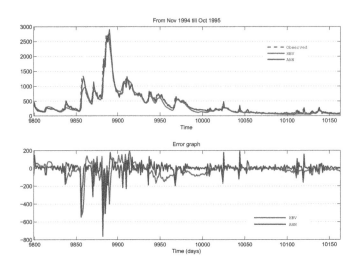

Figure 7.7: *Simulation results and error graph from HBV and ANN models for the year 1995*

Figure 7.6 shows the clear match between the time series shape of the HBV model and the measured runoff. No evident trend can be seen in the error graph, however, a clear bias (underestimation of the discharge) is seen on most low flows. It can be observed that the performance of the HBV and ANN models, in the plotted events, are not good in three sections of the hydrograph. The graph represents the beginning of the year 1993 and shows high complexity in the peak situations of this period, and all other errors can be seen on high flow for both models (above 260 m^3/s). A more detailed analysis shows that most of the ANN model errors are related with the steep changes of the hydrograph. On slow increasing phenomenas the problems appear only on the peaks or abrupt changes. These problems were also observed in other critical years (Figure 7.7).

Although high correlation makes the ANN model to give more weight to the discharge values than to the precipitation input, the different error measures show that there is no performance reduction using the precipitation for one time step. Comparing these models with the combined modelling approaches (Table 7.2), we can see that the he Nash Coefficient (CoE) of these models shows no clear difference between error corrector and the committee models; however, its their RMSE is better than the ANN and HBV models. In this sense there is no significant difference between the committee model with the linear and non-linear model combinations. Comparing the best models, of EC we can see that the best performance is achieved by the non-linear integration of models using Exogenous variables (HBV model results, past measures or corrected model results). In the CM model results no clear difference can be seen, and since the rainfall and runoff are considered non-linear in the multiple time step analysis, we will consider only the CM-ANN for extended lead times.

Table 7.2: *Performance of the CM and EC model types (1 day forecast).*

	CM-LM	CM-ANN	EC-LM	EC-ANN	EC-ANN-AR
RMSE	35.1913	35.2059	39.7591	31.3608	37.9378
CoE	0.9745	0.9745	0.9632	0.9797	0.9665
Cor	0.9873	0.9876	0.9816	0.9898	0.9832
NRMSE	15.9698	15.9765	19.171	14.2316	18.2927
MAE	20.9488	19.9325	23.1965	18.323	22.105
PERS	0.6808	0.6805	0.414	0.7465	0.4664
SSE	1404400	1405500	1797400	1115300	1636500

The performance results for one time step ahead shows the clear limitation of the HBV model in accuracy at simulating one time step compared with the neural networks and the integrated models (EC or CM). This is expected since it does not include the measured discharge at the previous time step, which has an important auto-correlation. However, the hydrograph generated by HBV model in general represents the rainfall runoff trend very well, specially for the most critical situations. Figure 7.6 and Figure 7.7 show the HBV and ANN models results, for the floods in 1993 and 1995, respectively.

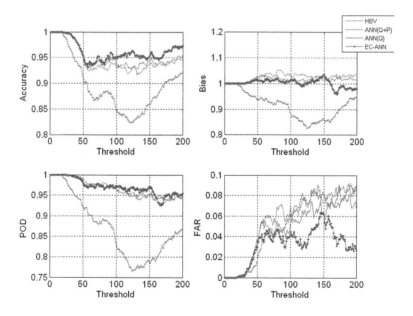

Figure 7.8: *Percentage of correct forecasts, POD, FAR and bias for the ANN(Q), ANN(Q+P), EC-ANN and HBV*

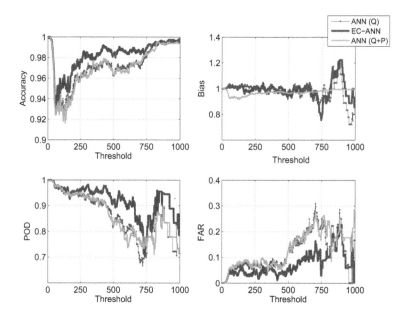

Figure 7.9: *Percentage of correct forecasts, POD, FAR and bias for the ANN(Q), ANN(Q+P) and EC-ANN*

Error corrector models were set up with different combinations of input variables in order to use as much information as possible: the HBV model past results, past error, past precipitation, and past internal states were used. All these experiments are set up following the average mutual information (AMI) and correlation analysis leading to the relevant input selection, as mentioned by Bowden et al. (2005a). The results show that the past errors in the HBV model are the most important variable, and, as it was expected, without them the models did not have acceptable performance. Further, the most accurate ANN model setup included 3 lags of the measured error, the 3 lagged measured discharges and the data from the 15 precipitation stations. It seems that for the first time step the performance of the committees is lower than that of autoregressive linear or non-linear (ANN) error corrector-based models (Table 7.2).

Figure 7.8 presents the skill scores error measures for different thresholds (for one day ahead prediction). One can see clearly how the performance of HBV model varies and has minimum in a low flow region; from 40 to 200 m^3/s. In contrast to DDMs, which exhibit the high probability of detecting this low flow (95%). Figure 7.9 shows an extended evaluation of the forecasting capabilities for extended thresholds only for the ANN(Q), ANN(Q+P) and the EC-ANN. We can see that although the EC has the highest accuracy, the ANN(Q+P) has less bias. On one time step, it would appear that the precipitation brings an important information to the bias in the high flows. This is the only forecast score that seems to show an advantage in the use of discharge in the forecasting ANN model that uses precipitation. If the forecast precipitation information is not accurate, this affects the performance of the HBV conceptual model and does not influence significantly the ANN models. In previous studies (Hidayat, 2007), the error of the conceptual model in the Meuse river basin may increase by at least 20% while the ANN might reach only 0.5%. This studies were based on the inclusion of a proportional inclusion of random noise to the measured precipitation on three basins.

7.6.2   Results on multi step forecast

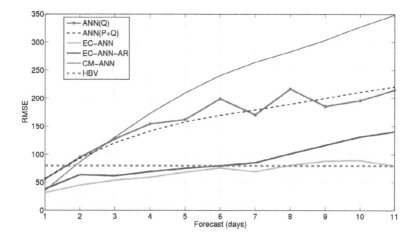

Figure 7.10: *Comparison of various models' performance in multiple time steps forecast*

The results of extended forecast to multiple time steps ahead is shown in Figure 7.10. As is can be seen the errors increase significantly after the second time step in the ANN model. Under the assumption of perfect precipitation and temperature forecasts, HBV model would always have the same performance. In this case the HBV model is better than ANNs and CM-ANN model after the second time step. However, the error corrector seems to reach even 7 days forecast. The EC-ANN-AR seems to have less accuracy than the EC-ANN that uses exogenous variables like the HBV model results and HBV previous corrected model.

The ANN(Q) has less accuracy than the ANN(Q+P) after the second time step. High flows that dominate the errors on extended lead times are probably influenced by the precipitation information. In addition to this, it can be mentioned that including precipitation as part of the input helps on the correction of the timing errors commonly present in ANN forecasting models (Abrahart et al., 2007), which was also observed partially in our results.

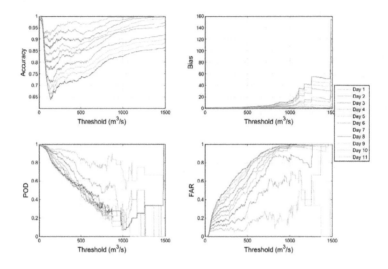

Figure 7.11: *Comparison of various models' performance in high flows for multiple time steps forecast*

Figure 7.11 shows the different skill score measure for the ANN(Q+P). We can see that POD scores are critical in almost all forecast (1 to 11 days) for values around the mean value of the river discharge (250 m^3/s). Above $100m^3/s$ the Bias, POD and FAR show that almost all forecast have variable performance, and therefore 10 days forecast could have better forecast than 9 days, or any combination of forecasts.

7.7 Conclusions

This chapter presented two contrasting approaches in hybrid modelling: parallel architecture (ensembles with linear and non-linear combining schemes), and the sequential one with the error correctors applied to the main model forecasts. This was done in order to assess the performance, advantages and disadvantages of these techniques for multiple time steps ahead forecasting.

The performance of ANN models under one and multiple time steps is quite different, and therefore it was important to explore the influence of other inputs on the ANN performance. As expected, the forecast results deteriorate when lead times are increased. In fact the ANN models do only have one maximum 2 days lead time forecast better than the HBV conceptual model. The error corrector models outperform single ANN models. For short time steps hybrid approaches of ANN and HBV model corrected by the non-linear error corrector show to be the most accurate. This hybrid approach also does not have high error variability between different time steps as it is the case for all other models.

Inputs fed into the models contain information that in operational real-life situations may not be available. In this sense it is expected that under inaccurate precipitation and temperature forecasts the hybrid schemes or even single ANN models will be more robust than single error corrected conceptual hydrological models like the HBV; this however needs more research.

ANN models for forecasting may be misleading since it appears that the ANN model outperforms the process-based hydrological model in the first time steps (simulation time), however, in extended lead time this will not represent the same difference in accuracy. This lead time is highly relative to the performance of the precipitation forecast information used by the conceptual hydrological model.

The implications of these results in operational systems require an extended analysis of all the different error contributions in the model. Such analysis has been performed for one time step forecasts (Corzo et al., 2009a), throwing some light on how the sub-basins contribute to the overall flow and how the data-driven models can interact. Further studies would be needed to analyse the sensitivity of the model to the quality of the weather forecast, to the errors in the routing model, and to the other spatially distributed variables.

MODULAR MODELS BASED ON CLUSTERING: APPLICATION TO PRECIPITATION DOWNSCALING FROM GENERAL CIRCULATION MODELS

It has been shown in the previous chapters that the application of modular models in the context of rainfall-runoff forecast modelling leads to increase in accuracy. This chapter explores the use of the modular approach technique in yet another area related to forecasting downscaling of information from General circulation models (GCM) and/or numerical weather prediction models (NWP). The use of a version of modular approach, fuzzy committee model (Solomatine, 2006; Solomatine and Price, 2004) is used along with the automatic clustering techniques, and compared with statistical downscaling models and single neural network models. This chapter is mainly based on the publication presented by Corzo et al. (2009b)

8.1   Introduction

Global Circulation Models and downscaling

In the last decades multiple Global Circulation Models (GCM), able to simulate the interactions among the atmosphere, oceans and surface, have been developed. to assist in the analysis of probable future weather scenarios (Wilby and Wigley, 1997; Wilby et al., 1998). They are used for studying the dynamics of the weather and climate system and the projections of future climate (Houghton, 1996). GCMs demonstrated significant skills (accuracy) at the large scales and they incorporate a large portion of the complexity of the global system, but they are unable to represent local sub-grid scale features and dynamics (Wigley and Raper, 1992; Wilby and Wigley, 1997).

Many limitations related to data-assimilation, quality of models and computational problems still do not allow for building models with grid sizes small enough to be able to forecast the weather variable at catchment (hydrological units) scale for the whole earth. A solution is to use so-called *downscaling*: mapping the variables (precipitation, temperature) predicted by GCMs and attributed with the large grid cells, to smaller cells or certain points of interest (often, existing gauges), from which the information could be fed to regional hydrological models. In this regard, there are two main challenges.

The first challenge relates to the number of output variables that result from the climate model (e.g. 26), which needs to be combined and downscaled in order to have an adequate model representation of the few variables required in the hydrological model (e.g., P, Et, T). The second challenge is linked to the uncertainty in the results of GCMs, linked with different quite hypothetical scenarios of population growth and overall gas emissions (Khan et al., 2006). For the former problem, downscaling of precipitation and temperature variables have been explored using standard statistical and computational intelligence models. It has been argued that the statistical models fail in capturing the seasonality (Masoud et al., 2008).

This problem is quite well known in computational intelligence where artificial neural network (ANN) models may not perform well when increased seasonal complexity is observed. Some researches show that the performance of the data-driven models like ANN is better than that of traditional statistical approaches (Liu et al., 2008; Yonas and Coulibaly, 2006). Nevertheless, some special cases in the past showed the failure of ANN in seasonal or problems with the changing regime: e.g., Wilby et al. (1998) found that ANN models perform poorly due to failures under specific conditions (e.g. wet days).

8.2   Fuzzy committee

In our studies presented in Chapter 3 and 5, we found that the modular modelling approach commonly achieve equal or better performance than single overall data-driven models. In this chapter a modular model belonging to class MM1, presented in Chapter 3, is applied to the problem of downscaling information form GCM models. One of the limitations in the use of multiple (modular) models is possible incompatibility at the boundary between the input sub-spaces handled by different models this contradicts the typically continuous character of physical phenomenon that is characterized by a single state vector (see Appendix A. One of the ways to soften this problem is to use the so called fuzzy committee (Solomatine, 2006; Solomatine and Price, 2004) that smoothens the transition between the models. This approach was applied to modular conceptual hydrological models (Fenicia et al., 2007), and now is used in this chapter as well.

This study presents the use of a modular and committee modelling as a complementary technology in statistical downscaling. The fuzzy committee machine model (FZCM) presented in this research can be seen also as version of the modular model MM1 presented in an analysis of flow condition based on clustering techniques in Chapter 3.

8.3   Case study: Beles River Basin, Ethiopia

Data from GCM

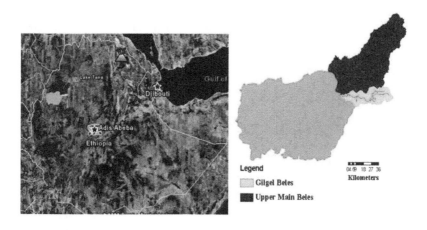

Figure 8.1: *Tana and Beles Basin Map (SMEC, 2008)*

Reanalysis data of the National centre for environmental prediction, (NCEP, USA), and the measurements of a precipitation gauge inside the Beles basin (Ethiopia) have been used. Re-analysis data are fine resolution grid data which combine observations with simulated data from numerical models, through data assimilation (more on this see IPCC-TGCIA, Barrow et al. 2000). The NCEP data was re-grid to conform the grid system of scenario HadCM3 (Canadian Climate Change Scenarios Network, CCCSN). NCEP re-analysis II predictor data files were downloaded from the Canadian Institute for climate studies (CICS) website, although the reanalysis data is provided by the NOAA/OAR/ESRL PSD, Boulder, Colorado, USA. The atmospheric part of the model has a grid of 2.50 latitude by 3.75 longitude (i.e. gives a resolution of approximately $300km$). The predictor variables available for this study are shown in Table 8.1. All predictors, with the exception of wind direction, have been normalized with respect to the 1961-1990 mean and standard deviation. The data used in this study covers predictor variables from 1961 till 2001.

Table 8.1: *List of 26 predictor variables derived from African window for the grid on the study area. p5 and p8 stands for the location at 500 hpa and 850 hpa respectively*

Variable No.	Predictor Variable	Description	Variable No.	Predictor Variable	Description
1	Mslpaf	Mean sea level pressure	14	p500af	500 hpa geopotential height
2	p5_ faf	Surface air flow strength	15	p850af	850 hpa geopotential height
3	p5_ uaf	Surface zonal velocity	16	p_ faf	Airflow strength
4	p5_ vaf	Surface meridional velocity	17	p_ thaf	Wind direction
5	p5_ zaf	Surface vorticity	18	p_ uaf	Zonal velocity
6	p5thaf	Surface wind direction	19	p_ vaf	Meridional velocity
7	p5zhaf	Surface divergence	20	p_ zaf	Vorticity
8	p8_ faf	Geostatic air flow velocity	21	p_ zhaf	Divergence
9	p8_ uaf	Zonal velocity	22	r500af	Relative humidity
10	P8_ vaf	Meridional velocity	23	r850af	Relative humidity
11	P8_ zaf	Voriticity	24	rhumaf	Near surface relative humidity
12	p8thaf	Wind direction	25	shumaf	Surface specific humidity
13	p8zhaf	Divergence	26	tempaf	Tempreature at 2 m

8.4   Beles River Basin

The Beles basin is one of the major sub-basins of upper Blue Nile. The main stem of the Beles River originates on the face of the escarpment across the divide to the west of the south-western portion of Lake Tana. It then flows on in a westerly direction and enters into the Blue Nile just before it crosses the Ethiopia-Sudan frontier. It is the only major right bank tributary of Blue Nile. The Beles basin covers an area of about 14,000 km^2 and geographically it extends from 10° 56' to 12° N latitude and 35° 12' to 37° E longitude.

The basin has two gauged sub-catchments, Main Beles and Giligile Beles that have a size of 3474 km^2 and 675 km^2 respectively. In particular the main focus of this study is the gauged part of the basin called Upper Beles sub-basin. Figure 8.1 shows its location on the Blue Nile basin map, along with the other two basins Tana and Beles basins (Upper main Beles and Giligile Beles sub-basins). The measurements from the station located at Pawe are used in this study. The NCEP re-analysis data is adapted from measured information for the period between 1987 and 2001, and therefore is used as the source of information for building the models described in this study. Period from 1987 to 1996 is used for training and from 1997 till 2001 for validation.

8.5 Methodology

In order to select the most representative variables for the modelling process, correlation analysis and average mutual information (AMI) analysis using all the 26 variables were performed. Figure 8.2 shows on the x-axis the variable number from Table 8.1, and on the y-axis (a) the maximum cross correlation found and cross correlation at no lag, (b) the maximum AMI and (c) the lags for which the maximum AMI or cross correlation were found. Cross-correlation and AMI analyses give similar results. It appears that the use of lagged information for the models, in order to increase the performance, not only has no clear physical interpretation, but also in terms of cross correlation and AMI is not really significant. This can be seen in the Figure 8.2(c), where the most representative variables according to AMI and correlation values are above 5 days, which is quite far from a reasonable value for a physical phenomena relation.

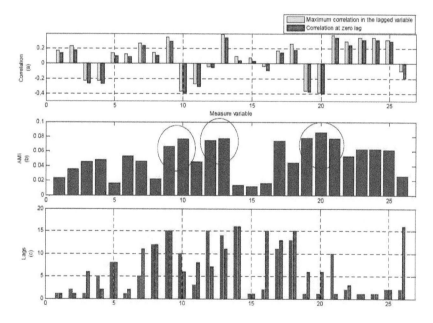

Figure 8.2: *AMI and correlation values obtained from the variables presented in Table 8.1 Blue colour corresponds to average mutual information and green to correlation coefficient values*

The three types of models were built:

- single overall ANN model;

- modular models based on cluster analysis (in nomenclature of Chapter 3, MM1);

- fuzzy committee model (FZCM).

A short description of the models is presented below.

8.5.1 ANN model setup

For downscaling of the weather information, it is reasonable to use a time-lagged feed forward neural network (TLFN) (Yonas and Coulibaly, 2006). However, the weather models are setup in grids that are hardly autocorrelated, and the dynamics of the cell can not be clearly related with previous situations in a daily time step analysis. Although, as it can be seen in the Figure 8.2 , there is an increase in the correlation with some time steps, this is not significant. Since the purpose of this analysis is to compare data-driven models dealing with relatively clear physical variables, no lags are used in the process. If the use of lags is desirable the correlation and AMI analyses described in Chapter 4 can be used for their identification (Bowden et al., 2005a; Corzo and Solomatine, 2007b; Luk et al., 2000)

For the different ANNs elaborated in this study a number of common features were defined. The first one is the neural network training process (calibration), which was performed using the Levenberg-Marquardt algorithm. The parameters of training were: the learning rate of 0.1, momentum term of 0.01, with 150 epochs. The number of epochs was defined after a visualization of the error performance in higher number of trials with different epochs. The networks are defined to have only one hidden layer, with multiple nodes (optimized) with sigmoid functions on it. The output layer as well contains a sigmoid function. The intention of it is to avoid negative values which will be limited the normalization (0.1-0.9), giving some extrapolation capacity to the network (Hettiarachchi et al., 2005). All inputs are normalized before being input to the network. The number of hidden nodes in the hidden layer was optimized and appeared to be different for different modules of the modular model. This is explained by the fact that the models represent different processes.

8.5.2 Committee and modular models

Weather variables are characterized by variability and complex interactions which are quite difficult for data-driven models to capture. For examples, in Figure 8.3 the positive divergence seems to lead to precipitation, but at the same time some isolated negative and positive patterns are present in situation where no precipitation is measured. This shows the high complexity in identifying the proper variables for downscaling models, as well as the possible need to separate situations with and without precipitation. In the considered case study, a simplified approached using k-means clustering has been applied.

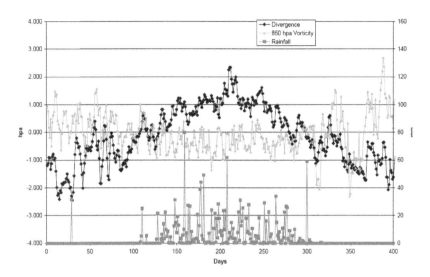

Figure 8.3: *Comparison of the results of the fuzzy clustering for the first 500 samples.*

Clustering input vectors (MM1)

This modular scheme presented as MM1 in Chapter 3 is used. To identify the clusters the k-means algorithm (Hartigan and Wong, 1979) with two and three clusters was tested. After the visual analysis of the clusters found, the number of cluster to be used was set to 2, which corresponds to the wet and dry environmental conditions. The best distance metric (Euclidean distance) for the clustering scheme was evaluated by a visual inspection, through an iterative process.

Learning classifier (LC-S)

Learning the cluster pattern is required in order to allow the new data to be classified into one of the clusters (for operation purposes). The algorithm has to be applied to build a classifier that would classify the new examples according to the cluster identified by the k-means algorithm. For this purpose we tested the three classification algorithms mentioned in Chapter 3 and the regression tree algorithm was chosen for its accuracy. During testing and operation phase the trained classifier attributes the new examples to one of the two trained ANN models.

8.5.3 Fuzzy committee machine

For combining models, the fuzzy committee approach has been used (Solomatine, 2006). As mentioned in Chapter 2, committee machine models are a generalization of the modular models and other separation and integration techniques in computational science. In the method used, the model combiner uses fuzzy membership of the samples used as inputs to the models. This is done in order to achieve an improvement in the transition between models. The most common approach is to use a fuzzy clustering technique, in this study we used fuzzy c-means clustering (Pal and Bezdek, 1995). All the samples in the input belong to a cluster with some degree of membership. To make the split a threshold has to be defined. One option to do this is presented in Figure 8.4, where a membership degree can be assigned to an input vector or sample based on a percentage of half of the Euclidean distance between cluster centres.

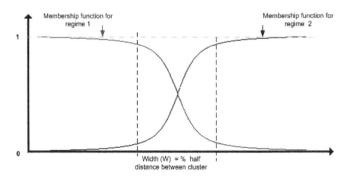

Figure 8.4: *General representation of two membership functions for a transition of regimes*

The main purpose of the fuzzy membership is to improve samples where it is not clear to which cluster (model) they should belong due to the relative similar distance to two or more clusters. Using modular models with a fuzzy inference model (Abraham, 2001; Pal and Bezdek, 1995) allows for weighting the results of each modular model according to the degree of belonging of the input vector to one of another cluster.

To create a membership function the fuzzy c-means clustering technique was used. As is shown in Figure 8.5, the training process is based on samples classified to a group after applying a threshold of the degree of membership found. The fuzzy inference model used in this study is trained to find the weights that complement the result required in training to obtain the target for each cluster. The weight was defined only on the boundary above 90% of half of the distance between cluster centers. For points belonging to such region equation 8.1 was assumed. Based on this the weight for that region can be defined as the relationship between the rainfall and the modular model results, determined in the fuzzy c-means clustering and reproduced by a surrogate ANN model.

$$MM_1 \cdot W_1 + MM_2 \cdot W_2 = R \qquad (8.1)$$

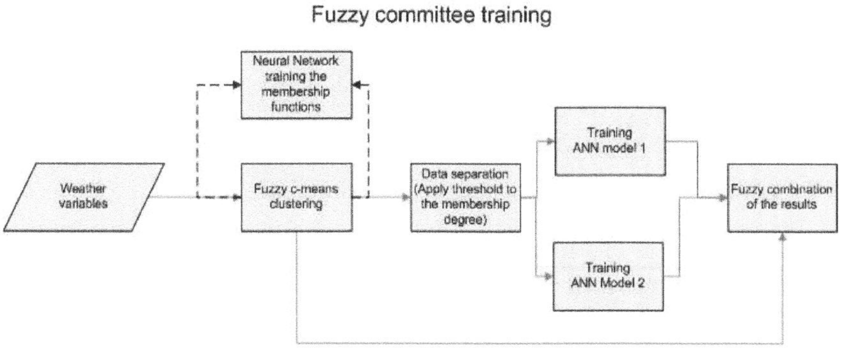

Fuzzy committee training

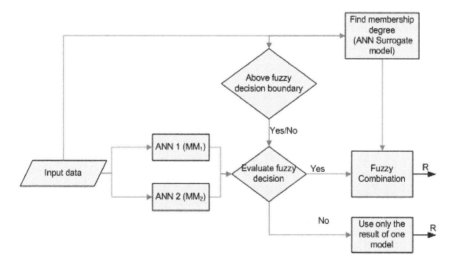

Figure 8.5: *Fuzzy committee model training scheme*

where $W_{1,2}$ represent the weight for cluster one and two respectively. represents the modular model results for each cluster, and R is the rainfall; only used when the distance to the cluster is above the 90% of half of the distance between centers (Figure 8.6).

Figure 8.6: *Fuzzy committee model in operation*

8.6   Results

Several models were compared in terms of their RMSE and correlation coefficient (Table 8.2). The ANN models built on full data sets used (a) the 26 variables with no lag and (b) less variables due to the principal component analysis (PCA) transformation. The results show that the performance of this transformation allows for obtaining better results than the SDSM model and Time lag feed forward neural network (TLFN). These models were elaborated following a 10 fold cross validation with the training data and selecting the best model for the validation.

Table 8.2: *Comparison of committee and non-committee model performance*

	ANN (26 Var)	ANN (PCA)	MM (No PCA)	FCM (No PCA)	TLFN	SDSM
Correlation coefficient (CC)	0.389	0.393	0.42	0.42	0.4	0.39
RMSE	10.01	9.96	9.86	9.85	10.3	10.1

Modular models outperform ANN models with PCA transformation in terms of RMSE and correlation coefficient. The modular model was set up based on k-means clustering using Euclidean distance. The results of this clustering uses average mutual information to select the input variables for each one of the modular models. The variables plotted in Figure 8.7 are the ones identified with the maximum mutual information between the precipitation and the different predictors (from left to right, 10 Meridional velocity at 850 hpa, 12 wind direction at 850 hpa, 13 Divergence at 850 hpa, 19 Meridional velocity, 20 vorticity, 21 Divergence).

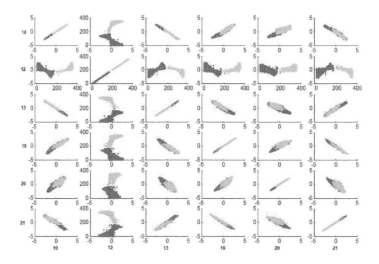

Figure 8.7: *Scatter plots of variables more correlated in the cluster selected for MM1.*

The cluster of variables show a clear separation between the data on wind direction (row and column 2 on figure 8.7) compared with all the other variables. However, for the other input dimensions there is no clear pattern to separate the phenomena and therefore is no clear difference in their grouping or classification.

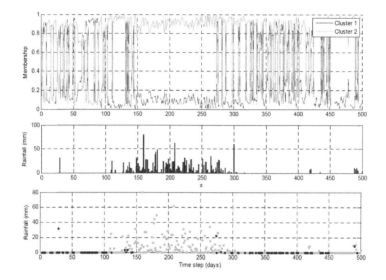

Figure 8.8: *Comparison of the results of the fuzzy clustering for the first 500 samples.*

The FZCM model results show that there is no significant transition regions, and that could be the reason why the performance of FZCM is so similar to that of the MM. For the fuzzy clustering the degree of membership of the two clusters obtained by using fuzzy c-means clustering are plotted in figure 8.8 . These clusters are created with Euclidean distance and using all the input variables available (no rainfall is considered). It is possible to see that the degree of membership for the precipitation is much higher for cluster 2 than for cluster one (for all the precipitation phenomena). However, for regions where almost no precipitation is present, cluster one has no clear or visual pattern. In this sense the transition between phenomena seems to be quite sharp. Other experiments where carried out using the rainfall variable to have a first split of data, but as it can bee seen in the Figure 8.8 rainfall with lower values that 1 or 2 are present along all the different seasons.

To analyze more the common region in the combination of the modular models a probability density function related to model verification was plotted (Figure 8.9). The full modular model (Ensemble of MM) results and each one of its clusters (denoted as P MM1 and P MM2) do not reproduce values above 12 mm. Also the modular model 1 is trained on samples that follow almost linear distribution with the low values (0-2.5 mm), and for the modular model 1 a mixed region for low and high values a nonlinear distribution with a clear asymptotic behaviour towards 12 mm. It appears that the separation is still not perfect and that the model that represents critical situations is quite limited. This problem is present in most of the other models as well, and seems one of the reasons of the complexity in downscaling precipitation.

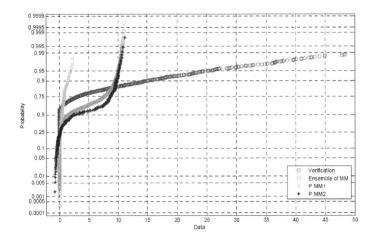

Figure 8.9: *Comparison of probability density function for the modular models and its ensemble.*

8.7   Conclusions

This chapter presented the comparison of using time lag neural networks (TLFN), artificial neural networks, modular models and fuzzy committee models. These experiments are conducted for exploring the performance of these models in downscaling GCM data to gauge precipitation data in the Beles basin in Ethiopia. The experiments were set up to compare the use of inputs without transformation in a modular model approach, and using lagged and not lagged input variables. The results show a clear improvement in the use of a PCA transformation over processing with input lags using maximum sensitivity (TLFN) and ANN using no lagged variables and input reduction based on AMI and correlation analyses. It appears that concentrating efforts on splitting input information is less significant than transforming input variables. However, more effort research needs to be done in order to evaluate MM with pre-processing information.

For the considered case, the fuzzy committee models did not add significant improvement over the modular model since no clear transitions between the clusters where identified. The modular models and fuzzy committee models have, however, higher performance than the one overall model covering the whole phenomena.

Based on the analysis of the probability distribution functions of the modular models and the measured data, it is possible to see that there are still regions where the data attributed to modular models were not split well, and further improvements with physical knowledge and expert rules may help. Also the use of fuzzy committee models to combine the specialized models requires better optimization of the shape of transitions areas in the input variables domain.

CONCLUSIONS AND RECOMMENDATIONS

This study deals with defining and exploring the principles and methods of hybrid modelling applied to the problem of hydrological forecasting. Hybrid modelling involves two or more modelling paradigms, and the two paradigms considered were hydrological process-based modelling, and data-driven modelling. They were combined using the principles of modularization, as well as parallel and complementary modelling. It has been shown how these two modelling paradigms form hybrid models, and can interact and complement each other in flow forecasting. In the framework of the major case study, the developed models were tested as software components into the operational hydrological forecasting system for the Meuse river basin, based on the Delft/FEWS platform and the HBV model.

The following sections address particular important parts and concepts of the study.

9.1 Hybrid modelling

Three ways of integrating the process-based approach to hydrological modelling (and in general, hydrological information) with the data-driven approach to modelling were identified: a) incorporation of the process-based information into data-driven models (P2D); b) incorporation of the data-driven techniques into process-based models (D2P); and c) parallel and serial architectures (DPPS). These ways of building hybrid models can also be interpreted as the general classes of hybrid models, on this basis a general hybrid modelling framework and specific architectures were suggested. A number of publications related to each of these classes have been reviewed and certain knowledge gaps identified. In building hybrid models, the principle of modularity (i.e. identifying the sub-processes and building for them separate models, possibly belonging to different paradigms) was found to be important and productive.

9.2   Modular modelling

Three main considered principles of models modularization are: temporal, processes-based, and spatial.

Experiments with *temporal and process-based modularization* were carried out on different types of catchments (Bagmati in Nepal, Sieve in Italy, and Brue in the UK). For this, the P2D approach to building hybrid models (incorporation of hydrological concepts into data-driven models) was employed. It was described and developed in Chapters 3, and tested in Chapter 5. Instead of training a global data-driven model on the whole data set, the training set is partitioned into several subsets, and a number of local models, each responsible for a region of the input space, were built. This principle was tested on a problem of modelling flow, separating the base- and excess flow. Three different partitioning schemes were employed; based: on clustering, on a traditional baseflow separation method (which was however updated and improved to allow for algorithmic implementation), and on using the hydrological process filter (which was optimized by the genetic algorithm GA and Global Pattern Search, with the higher performance of the latter).

The use of domain knowledge incorporated in the algorithms for separation of the base flow proved to be effective. Since most of such algorithms cannot be directly used in operation (since they require the future values of flow), they have to be replicated by surrogate classifier models, and it was shown that this approach can be successfully implemented. Several classifiers used in this role were compared in accuracy but the difference appeared to be marginal (The linear classifier was, however, the best). Partitioning the data by clustering in the input space leads to less accurate models when compared to those based on knowledge-based partitioning (flow separation). However, clustering-based partitioning is simple, is not sensitive to the algorithm parameters and can be used as a complementary tool.

The ensemble combination of the modular models in a fuzzy committee machine (where the areas of the inputs space covered by different models were only partly overlapping) was considered as well. It showed that such an approach will not always be necessary since in many problems there is no "soft" transition between regimes characterized by the data clusters (this was confirmed in the downscaling case study as well). Appropriate optimization of the fuzzy committees used may lead to better results. In general, it can be concluded that the modular approach for hydrological forecasting, especially the one involving the domain knowledge in partitioning the data and in building local specialized ANN models. The optimized overall model structure ensures accurate representation of the sub-processes constituting a complex natural phenomenon. Still more research on the possible uncertainty introduced by the hybrid approach needs to be done.

The principle of *spatial modularization* was developed and tested in Chapter 6. The main case study was the process-based semi-distributed rainfall-runoff (HBV) model of the Meuse river basin. The separate models were built for

different spatial subsystems (sub-regions). Spatial modularity was tested with
the two schemes of introducing data-driven model components into a semi-
distributed process based rainfall-runoff model (they belong to D2P class of
models). The first scheme explored the replacement of HBV sub-basin models
by ANN-MLP models using several scenarios. The results show that this approach improves the discharge simulation both in terms of reducing the RMSE
and increasing the model efficiency. The improvement was mainly observed for
the summer periods for low flows. The second scheme used the replacement
of the routing model (combining the individual sub-basin models) by an ANN,
and lead to higher gains in terms of the overall model accuracy than the first
scheme. Nevertheless, it is important to stress that this latter scheme does
not only reproduce the flow, but also the noise in the system. Therefore, the
second scheme (model) should not be seen as a more accurate river routing model, but more a combination of models that may act as error corrector as well.
In general, it can be concluded that both spatial combination schemes have a
clear potential in improving the accuracy of the considered class of hydrological
models.

Recommendations: A natural next step for the Meuse case study, would
be to move from daily to hourly data, and from one-step ahead forecasts to
models that forecast the flow several time steps (hours) ahead.

To address the problem of limited and inaccurate data for most of the sub-
basins, which affects the models performance, it would be reasonable to use
autoregressive models which are not sensitive to the precipitation, temperature
and evapotranspiration.

Further improvements in model accuracy could be expected on the route of
the models towards "optimal hybridization", when the processes to be modelled would be optimally partitioned in terms of the hydrological concepts, time
and space, and the best modelling paradigm would be chosen for each of the
sub-models. A number of issues are still to be resolved: including the choice
of adequate criteria of optimality, the balance between expert knowledge and
optimization algorithms, the inclusion of spatially distributed weather forecasts
into models, and the detailed analysis of the influence of various sources of uncertainty, mainly associated with lack of data and inadequate model structures.

9.3   Downscaling with modular models

Statistical (data-driven) downscaling, due to the identified seasonal effects, was
considered to be an area where the modular approach could be useful. For a
case study in Ethiopia, the modular approach lead to an increase accuracy in
downscaling the precipitation in comparison to the "single-model" approach,
both data-driven and statistical. However, in downscaling the temperature
the effect of modularization was marginal. The latter can be explained by
the fact that the temperature is a more periodic variable than precipitation,
and its relatively slow transition between low and high values makes it a less

appropriate variable for driving modular models. For the considered case study, the use of *fuzzy committees* for better integration of sub-models did not bring improvement over the "clear-cut" modular models.

Recommendations: For data-driven downscaling, it is recommended to explicitly include more knowledge about the physical processes, thus increasing the hybridization of models. This concerns both the single-model and the modular approaches. The ways such knowledge can be represented and incorporated have to be investigated.

9.4   Parallel and serial modelling architectures

Along with the P2D and D2P approaches to building hybrid models, the parallel and serial (DPPS) approaches in hybrid modelling were identified as well. The parallel architecture is associated with ensembles formed with linear and non-linear combining schemes, and the serial (sequential) architecture – with complementary error correctors applied to the main model forecasts.

The multiple combinations of ensembles and error corrector models were tested. The committee models, employing ANN and the HBV models for the Meuse river basin, have similar performance as a model with an ANN error corrector with short lead times. In the Meuse case study the non-linear error corrector was found to be better than the linear error correctors. The results show that adding the error corrector improves the accuracy of the HBV for the lead times which are even higher than the concentration time. It appeared from experiments that a single ANN cannot produce accurate forecasts for lead times higher than the characteristic lag (travel) time of the particular river. These experiments were based on the assumption of perfect rainfall forecast (hindcasting), but can be extended for real forecasts. It general, it was shown that the limitations of the process-based models can be overcome by complementary error correcting data-driven models.

Recommendations: Since in the considered case study ideal forecasts were used, it is recommended to study the sensitivity of the models to the accuracy of the precipitation and temperature forecasts. It may be expected that the hybrid schemes or even single ANN models will be more robust than single error corrected conceptual hydrological models like the HBV, but to justify this more research is needed.

It would be also advisable to investigate the possibility of building adaptive serial hybrid models (error correctors), allowing for switching between different regimes (so that error correctors would be modular themselves), and to test these on various case studies.

9.5   Data-driven modelling

Although the use of data-driven models for flow forecasting has been extensively studied (ASCE, 2000a; Solomatine and Ostfeld, 2008), there are a number of issues that are recognized in many studies, but still receive relatively little attention: a) the problem of optimal selection of model variables; b) the problem of the optimal splitting of data into training, cross-validation and test sets; c) the sensitivity of ANNs to the random initialization of the weights; d) the relative effectiveness of using ensembles of differently initialized ANNs. All these issues were investigated in this thesis, and adequate procedures have been suggested. The sensitivity of the inputs due to data availability and with respect to different types of flow events was analyzed. Six different model types were evaluated with 12 differently generated data sets. Most experiments concerned the Ourthe river basin in Belgium (part of the Meuse river basin). The following could be concluded.

The variability of the ANN models performance due to the differences in the weights initialization needs to be always taken into account. However, in the considered case study, ANN models with the optimized set of the input-output variables are not influenced much by the different (random) initializations of their weights. The procedure of optimizing network model structures and the 10 fold cross validation was found to be useful and was used in other experiments in this study.

The correlation and AMI analyses give similar results in terms of choosing the best set of input variables.

Partitioning of data ensuring statistical similarity of training, cross-validation and test data sets is an important step in the process of building the data-driven models. The methodology presented in this study, namely a combination of statistical analysis and visual inspection; a simpler alternative to the fully optimal approach, but appeared to be effective. However, this method can be improved and further developed in a multi-objective optimization framework.

One of the issues considered was the use of the past discharge as one of the inputs. The use of discharge as input in the model obviously improves the average accuracy of a DDM due to its high autocorrelation with the output compared to that of precipitation. For a single one step forecast the precipitation input does not significantly reduce the mean model error over a long period of time, and for multi-time step forecast such an effect is relatively small.

However, considering only averaged indicators of model error, like RMSE, could be misleading. Different error measures were compared, and in cases where there is high autocorrelation, measures like the coefficient of efficiency are not useful for model performance comparison. Combination of error measures like RMSE, MAE and the time series performance index (e.g PERS) is a preferred option. Of course, precipitation is the driving variable during the start of extreme events and flash floods, and should be included as input. More over, one of the most common problems in iterative forecasting is the timing error in the forecast results, which can be reduced with the inclusion of the precipitation.

When the data-driven model is set up with the past discharges as inputs, the model becomes driven mainly by autocorrelation rather than by the simulation of the real ongoing process. Such model structure is different from that of a conceptual model, but there are also advantages in this: namely, the possibility to capture stochastic information implicitly in the measured data and the high accuracy obtained.

Among all the data-driven modelling techniques tested the ANN model had the best performance. Using an ensemble of differently initialized models leads to more accurate forecasts.

Recommendations: It appears that the performance of a DDM and the process-based model relates to the response time of the system (i.e. the lag time known from the physical description of the system). It is suggested 1) to explore how the implicit water accumulation capability of a sub-basin may help in building more accurate DDMs; 2) to develop and test on large basins like Meuse the distributed architectures of DDM – that would include precipitation measurements or forecasts attributed to different locations.

9.6 Conclusion in brief

This research presents a hybrid modelling framework where data-driven and conceptual process-based models work in a coordinated fashion, and their role and performance are optimized. Several principles of model hybridization and modularization – spatial, temporal and processes-based – are considered and explored on a number of real case studies. Advantages and disadvantages of various approaches for different lead times are evaluated and discussed. In the framework of one of the case studies, the developed models were incorporated as software components into operational hydrological forecasting system for Meuse river basin, implemented on the Delft/FEWS platform. This thesis contributes to hydrological flow forecasting and its findings, we hope, could be used in building more effective flood forecasting systems.

BIBLIOGRAPHY

Abbott, M. B.: The electronic encapsulation of knowledge in hydraulics, hydrology and water resources, Advances in Water Resources, 16, 21-39, 1993.

Abebe, A. J. and Price, R. K.: Managing uncertainty in hydrological models using complementary models, Hydrological Sciences Journal, 48, 679–692, 2003.

Abraham, A.: Neuro-Fuzzy Systems: State-of-the-Art Modeling Techniques, Proceedings of the 6th International Work-Conference on Artificial and Natural Neural Networks: Connectionist Models of Neurons, Learning Processes and Artificial Intelligence-Part I, pp. 269–276, 2001.

Abrahart, R. J. and See, L.: Comparing neural network and autoregressive moving average techniques for the provision of continous river flow forecasts in two contrasting catchments, Hydrological Processes, 14, 2157–2172, 2000.

Abrahart, R. J. and See, L.: Multi-model data fusion for river flow forecasting: an evaluation of six alternative methods based on two contrasting catchments, Hydrology and Earth System Sciences, 6, 655–670, 2002.

Abrahart, R. J., Heppenstall, A. J., and See, L. M.: Timing error correction procedure applied to neural network rainfallrunoff modelling, Hydrological Science Journal, 52, 414–431, 2007.

Abramson, M. A., Audet, C., and Dennis, J. E.: Generalized pattern searches with derivative information, Mathematical Programming, 100, 3–25, 2004.

Anctil, F. and Tapé, D. G.: An exploration of artificial neural network rainfall-runoff forecasting combined with wavelet decomposition, J. Environ. Eng. Sci, 3, 1, 2004.

Arnold, J. G. and Allen, P.: Validation of automated methods for estimating baseflow and groundwater recharge from stream flow records, Journal of American water resources association, 35, 411–424, 1999.

ASCE: Task Committee on Application of Artificial Neural Networks in Hydrology, Artificial Neural Networks in Hydrology. II:Hydrologic Application, Journal of Hydrologic Engineering, 5, 124–136, 2000a.

ASCE: Task Committee on Application of Artificial Neural Networks in Hydrology, Artificial Neural Networks in Hydrology. I: Preliminary Concepts, Journal of Hydrologic Engineering, 5, 115–123, 2000b.

Ashagrie, A. G., de Laat, P. J., de Wit, M. J., Tu, M., and Uhlenbrook, S.: Detecting the influence of land use changes on discharges and floods in the Meuse River Basin; the predictive power of a ninety-year rainfall-runoff relation?, Hydrology and Earth System Sciences, 10, 691–701, URL http://www.hydrol-earth-syst-sci.net/10/691/2006/, 2006.

Auda, G. and Kamel, M.: Modular neural network classifiers: A comparative study, Journal of Intelligent and robotic Systems, 21, 117–129, 1998.

Babovic, V. and Fuhrman, D. R.: Data assimilation of local error forecasts in a deterministic model, Numerical Methods in Fluids, 39, 887–918, 2002.

Barrow, E., Hulme, M., Semenov, M., and Brooks, R.: Climate change scenarios, Climate change, climate variability and agriculture in Europe: An integrated assessment, pp. 11–27, 2000.

Bartholmes, J. and Todini, E.: Coupling meteorological and hydrological models for flood forecasting, Hydrology and Earth System Sciences, 9, 333–346, 2005.

Berger, H. E. J.: Flow Forecasting for the River Meuse, Technische Universiteit Delft, 1992.

Bergström, S. and Forsman, A.: Development of a conceptual deterministic rainfall-runoff model, Nordic Hydrology, 4, 147–170, 1973.

Beven, K. J.: Rainfall-runoff modelling, The Primer, Jhon Wiley and Sons, Chichester, UK, 2003.

Booij, M. J.: Modelling the effects of spatial and temporal resolution of rainfall and basin model on extreme river discharge, Hydrological Sciences Journal, 47, 307–320, 2002.

Booij, M. J.: Impact of climate change on river flooding assessed with different spatial model resolutions, Journal of Hydrology, 303, 176–198, 2005.

Bowden, G. J., Dandy, G. C., and Maier, H. R.: Input determination for neural network models in water resources applications. Part 1background and methodology, Journal of Hydrology, 301, 75–92, 2005a.

Bowden, G. J., Dandy, G. C., and Maier, H. R.: Input determination for neural network models in water resources applications. Part 2. Case study: forecasting salinity in a river, Journal of Hydrology, 301, 93–107, 2005b.

Brath, A., Montanari, A., and Toth, E.: Neural networks and non-parametric methods for improving real-time flood forecasting through conceptual hydrological models, Hydrology and Earth System Sciences, 6, 627–639, 2002.

Bray, M. and Han, D.: Identification of support vector machines for runoff modelling, Journal of Hydroinformatics, 6, 265–280, 2004.

Breiman, L., Friedman, J., Olshen, R., and Stone, C.: Classification and Regression Trees, vol. 1, Chapman and Hall/CRC, 1984.

Broersen, P. and Weerts, A.: Automatic Error Correction of Rainfall-Runoff models in Flood Forecasting Systems, in: Instrumentation and Measurement Technology Conference, vol. Volume 2, pp. 363–368, IEEE, 2005.

Broersen, P. M. T.: The Quality of Models for ARMA, Processes, IEEE Transactions on Signal Processing, 46, 1749–1752, 1998.

Butts, M. B., Hoest-Madsen, J., and Regsgaard, J. C.: Hydrologic Forecasting, in: Encyclopedia of Physical Science and Technology, vol. 7, Acedemic Press, third edn., 2002.

Butts, M. B., Payne, J. T., Kristensen, M., and Madsen, H.: An evaluation of the impact of model structure on hydrological modelling uncertainty for streamflow simulation, Journal of Hydrology, 298, 222–241, 2004a.

Butts, M. B., Payne, J. T., and Overgaard, J.: Improving streamflow simulations and flood forecasting with multimodel ensemble, in: 6th International Conference on Hydroinformatics, edited by Liong, P. a. B., World Scientific Publishing, Singapur, 2004b.

Candela, J. Q., Girard, A., Larsen, J., and Rasmussen, C. E.: Propagation Of Uncertainty In Bayesian Kernel Models-Application To Multiple-Step Ahead Forecasting, in: The 2003 IEEE Conference on Acoustic, Speech and Signal Processing (ICASSP03), Hongkong, 2003.

Carpenter, T. M. and Georgakakos, K. P.: Impacts of parametric and radar rainfall uncertainty on the ensemble streamflow simulations of a distributed hydrologic model, Journal of Hydrology, 298, 202–221, 2004.

Chapman, T.: Modelling stream recession flows, Environmental Modelling and Software, 18, 683–692, 2003.

Chapman, T. G.: A comparison of algorithms for stream flow recession and baseflow separation, hydrological Processes, 13, 701–714, 1999.

Chen, J. and Adams, B. J.: Integration of artificial neural networks with conceptual models in rainfall-runoff modeling, Journal of Hydrology, 318, 232–249, 2006.

Cortes, C. and Vapnik, V.: Support-vector networks, Machine Learning, 20, 273–297, 1995.

Corzo, G. and Solomatine, D.: Multi-objective optimization of ANN hybrid committees based on hydrologic knowledge, in: Geophysical Research Abstracts, vol. 8, p. 09794, 2006a.

Corzo, G. and Solomatine, D.: Knowledge-based modularization and global optimization of artificial neural network models in hydrological forecasting, Journal of Neural Networks, 20, 528–536, 2007a.

Corzo, G. and Solomatine, D. P.: Numerical flow separation and committees of neural networks in streamflow forecasting, in: Geophysical Research Abstracts, vol. 8, p. 09732, 2006b.

Corzo, G., Solomatine, D., Hidayat, de Wit, M., Werner, M., Uhlenbrook, S., and Price, R.: Combining semi-distributed process-based and data-driven models in flow simulation: a case study of the Meuse river basin, Hydrology and Earth System Sciences Discussions, 6, 729–766, URL http://www.hydrol-earth-syst-sci-discuss.net/6/729/2009/, 2009a.

Corzo, G. A. and Solomatine, D. P.: A modular model approach in flow forecasting by neural networks, universities of Delft, Utrecht, Nijmegen, Twente and Wageningen, UNESCO-IHE, RIZA, ALTERRA, TNO-NITG and WL—Delft Hydraulics, 2005.

Corzo, G. A. and Solomatine, D. P.: Baseflow separation techniques for modular artificial neural networks modelling in flow forecasting, Hydrological Sciences Journal, 52, 491–507, 2007b.

Corzo, G. A., Siek, M., and Solomatine, D.: Modular data-driven hydrologic models with incorporated knowledge: neural networks and model trees, in: International congress IAHR, ASCE, Italy, 2007.

Corzo, G. A., Jonoski, A., Yimer, G., Xuan, Y., and Solomatine, D.: Downscaling global climate models using modular models and fuzzy committees, in: 8th international conference in hydroinformatics, edited by Liong, S.-Y., World Scientific Publishing Company, Concepcion Chile, 2009b.

Dawson, C. W., Harpham, C., Wilby, R. L., and Chen, Y.: Evaluation of artificial neural network techniques for flow forecasting in the River Yangtze, China, Hydrology and Earth System Sciences, 6, 619–626, 2002.

Dawson, C. W., See, L. M., Abrahart, R. J., Wilby, R. L., Shamseldin, A. Y., Anctil, F., Belbachir, A. N., Bowden, G., Dandy, G., and Lauzon, N.: A comparative study of artificial neural network techniques for river stage forecasting, 2005.

de Vos, N. and Rientjes, T.: Constraints of artificial neural networks for rainfall-runoff modelling: trade-offs in hydrological state representation and model evaluation, Constraints, 2, 365–415, 2005.

de Vos, N. J. and Rientjes, T.: Correction of timing errors of artificial neural network rainfallrunoff models, Hydroinformatics in Practice: Computational Intelligence and Technological Developments in Water Applications, 2007.

de Wit, M.: Van regen tot Maas, Veen Magazines, Diemen, rijkswaterstaat deltares edn., 2009.

de Wit, M. J. M., Peeters, H. A., Gastaud, P. H., Dewil, P., Maeghe, K., and Baumgart, J.: Floods in the Meuse basin: Event descriptions and an international view on ongoing measures, 2007a.

de Wit, M. J. M., van den Hurk, B., Warmerdam, P. M. M., Torfs, P., Roulin, E., and van Deursen, W. P. A.: Impact of climate change on low-flows in the river Meuse, 2007b.

Dibike, Y. B. and Abbott, M. B.: Application of artificial neural networks to the simulation of a two dimensional flow, Journal of Hydraulic Research/Journal de Recherches Hydraulique, 37, 435–446, 1999.

Dibike, Y. B. and Solomatine, D. P.: River Flow Forecasting Using Artificial Neural Networks, Physics and Chemistry of the Earth: B: Hydrology, Oceans and Atmosphere, 26, 1–7, 2001.

Diermansen, F.: Physically based modelling of rainfall-runoff processes, PhD Thesis - TuDelft, pp. 123–150, 2001.

Duband, D., Obled, C., and Rodriguez, J.: Unit hydrograph revisited: an alternate iterative approach to UH and effective precipitation identification, Journal of Hydrology(Amsterdam), 150, 115–149, 1993.

Duda, R. O.: Pattern Recognition for HCI, University of San Jose State, Web Notes., http://www.cs.princeton.edu/courses/archive/fall04/cos436/Duda/, 1996.

Ekhardt, K.: How to construct recursive digital filters for baseflow separation, Hydrological Processes, 19, 507–515, 2005.

Elshorbagy, A., Corzo, G., Srinivasulu, S., and Solomatine, D. .: Experimental investigation of the predictive capabilities of soft computing techniques in hydrology, CANSIM Series Report No. CAN-09-01, Centre for Advanced

Numerical simulation (CANSIM), Department of Civil & Geological Engineering, University of Saskatchewan, Saskatoon, SK, Canada, pp. 49, 2009a.

Elshorbagy, A., Corzo, G., Srinivasulu, S., and Solomatine, D.: Data driven techniques scrutinized: is there one better than the rest?, General Assembly of the European Geosciences Union, Vienna, Austria, April 19-24 (Oral presentation EGU2009-4282), 2009b.

Engel, B. A. and Kyoung, L. J.: WHAT (Web-Based Hydrograph Analysis Tool), 2005.

Fenicia, F., Solomatine, D., Savenije, H. H. G., and Matgen, P.: Soft combination of local models in a multi-objective framework, hydrological and earth systems science, 4, 91–123, 2007.

Fenicia, F., McDonnell, J., and Savenije, H.: Learning from model improvement: On the contribution of complementary data to process understanding, Water Resources Research, 44, 2008.

Fogelberg, S., Arheimer, B., Venohr, M., and Behrendt, H.: HBV modeling in several European countries, Proceedings of Nordic Hydrologic conference, 2004.

Friedman, J. H.: Multivariate Adaptive Regression Splines, The Annals of Statistics, 1991.

Georgakakos, K. P. and Krzysztofowicz, R.: Probabilistic and ensemble forecasting, Journal of Hydrology, 249, 1, 2001.

Georgakakos, K. P., Seo, D.-J., Gupta, H., Schaake, J., and Butts, M. B.: Towards the characterization of streamflow simulation uncertainty through multimodel ensembles, Journal of Hydrology, 298, 222–241, 2004.

Geva, A. B.: Non-stationary time-series prediction using fuzzy clustering, 1999.

Glemser, M. and Klein, U.: Hybrid Modelling and Analysis of Uncertain Data, IAPRS, XXXIII, 2000.

Hall, F.: Base-flow recessions review, Water Resources Research, 4, 973 983, 1968.

Hall, J. and Anderson, M.: Handling uncertainty in extreme or unrepeatable hydrological processesthe need for an alternative paradigm, Hydrological Processes, 16, 1867–1870, 2002.

Hartigan, J. A. and Wong, M. A.: A K-means clustering algorithm, JR Stat. Soc. Ser. C-Appl. Stat, 28, 100108, 1979.

Haykin, S.: Neural networks: a comprehensive foundation, Prentice Hall, second edn., 1999.

Hettiarachchi, P., Hall, M., and Minns, A.: The extrapolation of artificial neural networks for the modelling of rainfall-runoff relationships, Journal of Hydroinformatics, 7, 291–296, 2005.

Hidayat: Application of Data-Driven Techniques in Semi-Distributed Hydrological Modelling, Ph.D. thesis, 2007.

Houghton, J. T.: Climate Change 1995: The Science of Climate Change, Cambridge University Press, 1996.

Hu, T. S., LAM, K. C., and NG, S. T.: A modified neural network for improving river flow prediction, Hydrological Sciences Journal, 50, 299–318, 2005.

Jain, A. and Kumar, A.: Hybrid neural network models for hydrologic time series forecasting, Applied Soft Computing Journal, 7, 585–592, 2007.

Jain, A. and Srinivasulu, S.: Integrated approach to model decomposed flow hydrograph using artificial neural network and conceptual techniques, Journal of Hydrology, 317, 291–306, 2006.

Kachroo, R. and Liang, G.: River Flow Forecasting. Part 2. Algebraic Development of Linear Modelling Techniques, Journal of Hydrology JHYDA 7,, 133, 1992.

Kamp, R. G. and Savenije, H. H. G.: Hydrological model coupling with ANNs, Hydrology and Earth System Sciences, 11, 1869–1881, URL http://www.hydrol-earth-syst-sci.net/11/1869/2007/, 2007.

Khan, M., Coulibaly, P., and Dibike, Y.: Uncertainty analysis of statistical downscaling methods, Journal of Hydrology, 319, 357–382, 2006.

Kim, S., Yasuto, T., Sayama, T., and Takara, K.: Ensemble rainfall-runoff prediction with radar image extrapolation and its error structure, Annual Journal of Hydraulic Engineering, 50, 2006.

Kingston, G., Maier, H., and Lambert, M.: A probabilistic method for assisting knowledge extraction from artificial neural networks used for hydrological prediction, Mathematical and Computer Modelling, 44, 499–512, 2006.

Kitanidis, P. K. and Bras, R. L.: Real-Time Forecasting With a Conceptual Hydrologic Model: Analysis of Uncertainty, Water Resources Research, 16, 1025–1033, 1980.

Kohonen, T.: Self-organized formation of topologically correct feature maps, Biological Cybernetics, 43, 59–69, 1982.

Kuncheva, L. I.: Combining Pattern Classifiers: Methods and Algorithms, Wiley InterScience, 2004.

Lauzon, N., Anctil, F., and Baxter, C.: Clustering of heterogeneous precipitation fields for the assessment and possible improvement of lumped neural network models for streamflow forecasts, Hydrology and Earth System Sciences, 10, 485–494, 2006.

Leander, R. and Buishand, T. A.: Resampling of regional climate model output for the simulation of extreme river flows, 2007.

Leavesley, G. H., Markstrom, S., Restrepo, P., and Viger, R. J.: A modular approach to adressing model design, scale, and parameter estimation issues in distributed hydrological modelling, Hydrological Processes, 16, 173–187, 2002.

Leontaritis, I. and Billings, S.: Input-output parametric models for non-linear systems Part I: deterministic non-linear systems, International Journal of Control, 41, 303–328, 1985.

Linde, A. t., Hurkans, R., Aerts, J., and Dolman, H.: Comparing model performance of the HBV and VIC models in the Rhine basin, in: Symposium HS2004 at IUGG2007, IAHS, Perugia, 2007.

Lindström, G., Johansson, B., Persson, M., Gardelin, M., and Bergström, S.: Development and test of the distributed HBV-96 hydrological model, Journal of Hydrology, 201, 272–228, 1997.

Linsley, R. K., Kohler, M. A., and Paulhus, J. L.: Hydrology for engineers: New York, 1982.

Lischeid, G. and Uhlenbrook, S.: Checking a process-based catchment model by artificial neural networks, Hydrological Processes, 17, 2003.

Liu, X., Coulibaly, P., and Evora, N.: Comparison of data-driven methods for downscaling ensemble weather forecasts, Hydrology and Earth System Sciences, 12, 615–624, URL http://www.hydrol-earth-syst-sci.net/12/615/2008/, 2008.

Luk, K., Ball, J., and Sharma, A.: A study of optimal model lag and spatial inputs to artificial neural network for rainfall forecasting, Journal of Hydrology, 227, 56–65, 2000.

Madsen, H. and Skotner, C.: Adaptive state updating in real-time river flow forecasting; a combined filtering and error forecasting procedure, Journal of Hydrology, 308, 302–312, 2005.

Madsen, H., Butts, M. B., Khu, S. T., and Liong, S. Y.: Data assimilation in rainfall-runoff forecasting, Proc. of the 4th Int. Conference on Hydroinformatics, Iowa, USA, 2000.

Masoud, H., Philippe, G., Taha, B. M. J. O., Andr, and St, H.: Automated regression-based statistical downscaling tool, Environ. Model. Softw., 23, 813–834, 1344931, 2008.

McCuen, R. H.: Hydrologic Analysis and Design, Prentice Hall, Englewood Cliffs, Departament of Civil Engineering, University of Maryland, 1998.

Mitchell, T. M.: Machine Learning, McGraw-Hill, 1998.

Moore, B.: Special Issue: HYREX: the Hydrological Radar Experiment, Hydrology and Earth System Sciences, 4, 521–522, 2002.

Mor, J.: The Levenberg-Marquardt algorithm: implementation and theory, Lecture notes in mathematics, 630, 105–116, 1977.

Nash, J. E. and Sutcliffe, J. V.: River flow forecasting through conceptual models Part 1- A Discussion Principles, Journal of Hydrology, 10, 282–290, 1970.

Nilsson, N.: Learning machines: Foundations of trainable pattern-classifying Systems, McGraw-Hill., New York, 1965.

O'Connor, K.: River flow forecasting, River basin modelling for flood risk mitigation, p. 197, 2005.

Osherson, D., Weinstein, S., and Stoli, M.: Modular learning, Computational Neuroscience, MA:MIT Press, Cambridge, 1990.

Pal, N. and Bezdek, J.: On cluster validity for the fuzzy c-means model, IEEE Transactions on Fuzzy Systems, 3, 370–379, 1995.

Palmer, T. and Risnen, J.: Quantifying the risk of extrema seasonal precipitation events in changing climate, Nature, 415, 512–514, 2002.

Pan, T.-Y. and Wang, R.-y.: Using recurrent neural networks to reconstruct rainfall-runoff process, Hydrological Processes, 19, 3603–3619, 2005.

Parzen, E.: On Estimation of a Probability Density Function and Mode, The Annals of Mathematical Statistics, 33, 1065–1076, 1962.

Ponce, V. M., Lohani, A. K., and Scheyhing, C.: Analytical verification of Muskingum-Cunge routing, Journal of Hydrology, 174, 235 – 241, doi:DOI:10.1016/0022-1694(95)02765-3, URL http://www.sciencedirect.com/science/article/B6V6C-3VW173F-3/2/842b6f3ce5b836f6b48ebf8347774b88, 1996.

Price, R.: The growth and significance of hydroinformatics, River basin modelling for flood risk mitigation, p. 93, 2005.

Price, R. K.: Hydroinformatics for river flood management, Flood issues in contemporary water management, Kluwer, Amsterdam, 2000.

Price, R. K.: Phd Meeting discussions, Tech. rep., 2009.

Price, R. K., Ahamad, K., and Holtz, P.: Hydroinformatics concepts, vol. 44 of *Hydroinformatics tools for planning design and operation and rehabilitation of sewer systems*, Enviroment, 1996.

Pyle, D.: Data Preparationfor Data Mining, MorganKaufmann, USA, 1999.

Quinlan, J. R.: Bagging, Boosting, and C4.5, in: 13th International Conference on Artificial Intelligence, pp. 725–730, AAAI Press, Menlo Park, CA, Portland, OR, 1996.

Refsgaard, J. C.: Parameterisation, calibration and validation of distributed hydrological models, Journal of Hydrology, 198, 69-97, 1997.

Ronco, E. and Gawthrop, P.: Modular neural networks: a state of the art, Rapport Technique CSC-95026, Center of System and Control, University of Glasgow. http://www. mech. gla. ac. uk/control/report. html, 1995.

See, L. and Openshaw, S.: Applying soft computing approaches to river level forecasting, HYDROL SCI J, 44, 763–778, 1999.

See, L. and Openshaw, S.: A hybrid multi-model approach to river level forecasting, Hydrological Sciences Journal, 45, 523–535, 2000.

Seibert, J.: Estimation of parameter uncertainty in the HBV model, Nordic Hydrology, 28, 247–262, 1997.

Shamseldin, A. Y. and O'Connor, K. M.: A real-time combination method for the outputs of different rainfall-runoff models, Hydrological Sciences Journal, 44, 895–912, 1999.

Shamseldin, A. Y. and O'Connor, K. M.: A non-linear neural network technique for updating river flow forecasts, Hydrology and Earth System Sciences, 5, 577–597, 2001.

Shamseldin, A. Y., Abrahart, R. J., and See, L. M.: Neural network river discharge forecasters: an empirical investigation of hidden unit processing functions based on two different catchments, Neural Networks, 4, 2005.

Shrestha, D. L. and Solomatine, D.: Data-driven approaches for estimating uncertainty in rainfall-runoff modelling, Intl. J. River Basin Management, 6, 109–122, 2008.

Shrestha, I.: Conceptual and data-driven hydrological modelling of Bagmati river Basin, Nepal, Msc Thesis (UNESCO-IHE, Delft, the Netherlands), 2003.

Singh, V. and Frevert, D.: Mathematical models of small watershed hydrology and applications, Water Resources Pubns, 2002.

Sloto, R. A. and Cruise, M.: HYSEP: a computer program for streamflow hydrograph separation and analysis, Tech. rep., U.S. GEOLOGICAL SURVEY, 1996.

Solomatine, D.: Optimal modularization of learning models in forecasting environmental variables http://www.iemss.org/iemss2006/sessions/all.html., in: iEMSs 3rd Biennial Meeting: "Summit on Environmental Modelling and Software" ,, edited by A. Voinov, A. Jakeman, A. R., International Environmental Modelling and Software Society, Burlington, USA,, 2006.

Solomatine, D.: Data-driven modelling: machine learning, data mining and knowledge discovery, Lecture Notes, UNESCO-IHE, Delft, 2008.

Solomatine, D. and Ostfeld, A.: Data-driven modelling: some past experiences and new approaches, Journal of Hydroinformatics, 10, 3–22, 2008.

Solomatine, D. and Xue, Y.: M5 model trees compared to neural networks: application to flood forecasting in the upper reach of the Huai River in China, Journal of Hydrologic Engineering, 9, 491–501, 2004.

Solomatine, D. P.: Two strategies of adaptive cluster covering with descent and their comparison to other algorithms, J. Global Optimiz., 14, 55–78, 1999.

Solomatine, D. P.: Data-driven modeling and computational intelligence methods in hydrology, Encyclopedia of hydrological sciences, John Wiley & Sons, New York, 2005.

Solomatine, D. P. and Corzo, G. A.: Learning hydrologic flow separation algorithm and local ANN committee modelling, in: IJCNN '06. International Joint Conference on Neural Networks, pp. 5104 – 5109, Vancouver, 2006.

Solomatine, D. P. and Dulal, K. N.: Model tree as an alternative to neural network in rainfall-runoff modelling, Hydrological Science Journal, 48, 399411., 2003.

Solomatine, D. P. and Price, R. K.: Innovative approaches to flood forecasting using data driven and hybrid modelling, in: 6th International Conference on Hydroinformatics, edited by Liong, Phoon, and Babovic, pp. 1639–1646, World Scientific Publishing Company, Singapore, 2004.

Solomatine, D. P. and Siek, M. B.: Modular learning models in forecasting natural phenomena, Neural Networks, 19, 215–224, 2006.

Solomatine, D. P., Maskey, M., and Shrestha, D. L.: Instance-based learning compared to other data-driven methods in hydrological forecasting, Hydrological Processes, 22, 275–287.

Spath, H.: The Cluster Dissection and Analysis Theory FORTRAN Programs Examples, Prentice-Hall, Inc. Upper Saddle River, NJ, USA, 1985.

Specht, D. F.: Probabilistic neural networks, Neural Networks, 3, 109–118, 1990.

Stein, M.: Interpolation of Spatial Data: Some Theory for Kriging, Springer, 1999.

Sudheer, K. P.: Knowledge Extraction from Trained Neural Network River Flow Models, Journal of Hydrologic Engineering, 10, 264, 2005.

Sudheer, K. P. and Jain, A.: Explaining the internal behaviour of artificial neural network river flow models, Hydrological Processes, 18, 833–844, 2004.

Sudheer, K. P., Gosain, A. K., and Ramasastri, K. S.: A data-driven algorithm for constructing artificial neural network rainfall-runoff models, Hydrological Processes, 16, 1325–1330, 2002.

Toth, E. and Brath, A.: Flood Forecasting Using Artificial Neural Networks in Black-Box and Conceptual Rainfall-Runoff Modelling, in: Proceedings of the 1st Biennial Meeting of the iEMSs, vol. 20, 2002.

Toth, E., Brath, A., and Montanari, A.: Comparison of short-term rainfall prediction models for real-time flood forecasting, Journal of Hydrology, 239, 132–147, 2000.

Tsymbal, A., Pechenizkiy, M., and Cunningham, P.: Diversity in search strategies for ensemble feature selection, Information Fusion, 6, 83–98, 2005.

Tu, M., Hall, M., and Laat, P. d.: Detection of long-term changes in precipitation and discharge in the Meuse, GIS and remote sensing in Hydrology, Water resources and enviromental. - Proceedings of the international conference of the ICGRHWE held at the Three Gorges Dam, China, Septembre 2003, IAHS Publ. 289, 2004, 2004.

Tu, M., Hall, M. J., de Laat, P. J. M., and de Wit, M. J. M.: Extreme floods in the Meuse river over the past century: aggravated by land-use changes?, 2005.

Uhlenbrook, S., Frey, M., Leibundgut, C., and Maloszewski, P.: Hydrograph separations in a mesoscale mountainous basin at event and seasonal timescales, Water Resources Research, 38, 31.1–14, 2002.

Uhlenbrook, S., Roser, S., and Tilch, N.: Hydrological process representation at the meso-scale: the potential of a distributed, conceptual catchment model, Journal of Hydrology, 291, 278 – 296, doi:DOI:10.1016/j.jhydrol. 2003.12.038, URL http://www.sciencedirect.com/science/article/ B6V6C-4C1FDJB-1/2/55c6c5b068d2e5cb96d9d57b38bc8102, catchment modelling: Towards an improved representation of the hydrological processes in real-world model applications, 2004.

Valdés, J. and Bonhjam-Carter, G.: Time Dependent Neural Network Models for Detecting Changes of State in Earth and Planetary Processes, 2005.

van Deursen, W.: Calibration HBV model Meuse, Tech. rep., Carthago Consultancy, 2004.

Varoonchotikul, P.: Flood forecasting using artificial neural networks, Taylor and Francis Group, 2003.

Wahl, K. L. and Wahl, T. L.: Determining the flow of Comal Springs at New Braunfels, 1995.

Wang, W.: Sochasticity, nonlinearity and forecasting of streamflow processes, Ph.D. thesis, Technichal university of Delft, 2006.

Wang, W. and Ding, J.: Wavelet Network Model and Its Application to the Prediction of Hydrology, Science, 1, 67–71, 2003.

Wang, W., van Gelder, P., Vrijling, J. K., and Ma, J.: Forecasting daily streamflow using hybrid ANN models, Journal of Hydrology, 324, 383–399, 2006.

Wang, Y. and Witten, I.: Inducing model trees for continuous classes, Poster Papers of the 9th European Conference on Machine Learning (ECML 97), pp. 128–137, 1997.

Weerts, A. and El Serafy, G.: Particle filtering and ensemble Kalman filtering for state updating with hydrological conceptual rainfall-runoff models, Water Resources Research, 42, 9403, 2006.

Werner, M.: Spatial flood extent modelling A performance-based comparison, Ph.D. thesis, Delft University of Technology, The Netherlands, 2004.

Werner, M., Blazkova, S., and Petr, J.: Spatially distributed observations in constraining inundation modelling uncertainties, Hydrological Processes, 19, 3081–3096, 2005.

Wigley, T. M. L. and Raper, S. C. B.: Implications for climate and sea level of revised IPCC emissions scenarios, Nature, 357, 293–300, 1992.

Wilby, R. L. and Wigley, T. M. L.: Downscaling general circulation model output: a review of methods and limitations, Progress in Physical Geography, 21, 530, 1997.

Wilby, R. L., Wigley, T. M. L., Conway, D., Jones, P. D., Hewitson, B. C., Main, J., and Wilks, D. S.: Statistical downscaling of general circulation model output: A comparison of methods, Water Resources Research, 34, 2995–3008, 1998.

Wilby, R. L., Abrahart, R. J., and Dawson, C. W.: Detection of conceptual model rainfallrunoff processes inside an artificial neural network, Hydrological Sciences Journal, 48, 163–181, 2003.

Witten, I. H. and Frank, E.: Data Mining: Practical Machine Learning Tools and Techniques with Java Implementations, Morgan Kaufmann, 2000.

WMO: Simulated real time intercomparison of hydrological models, Tech. rep.

Xiaokun, Z., Weile, Z., and Deren, G.: A new global optimizing method and fast converge algorithm fornonlinear systems, Circuits and Systems, 1991. Conference Proceedings, China., 1991 International Conference on, pp. 924–927, 1991.

Xiong, L., Shamseldin, A. Y., and O'Connor, K. M.: A non-linear combination of the forecasts of rainfall-runoff models by the first-order Takagi-Sugeno fuzzy system, Journal of Hydrology, 245, 196–217, 2001.

Yonas, B. D. and Coulibaly, P.: Temporal neural networks for downscaling climate variability and extremes, Neural Netw., 19, 135–144, 1145899, 2006.

Young-Oh, K., DaeIl, J., and Hwan, K. I.: Combining Rainfall-Runoff Model Outputs for Improving Ensemble Streamflow Prediction, Journal of Hydraulic Engineering, 11, 578–588, 2006.

Zhang, B. and Govindaraju, R. S., eds.: Modular Neural Networks for Watershed Runoff, Artificial Neural Netwroks in Hydrology, R.S. Govindarju, and A. Ramachandra Rao (eds.), 2000a.

Zhang, B. and Govindaraju, R. S.: Prediction of watershed runoff using Bayesian concepts and modular neural networks, Water Resources Research, 36, 753–762, 2000b.

STATE-SPACE TO INPUT-OUTPUT TRANSFORMATION

A.1 State-space and input-output models

From the data-driven point of view systems work only with observed input-output data pairs, and do not contemplate explicitly important unobservable variables like internal states, z(t), which are not commonly available, or cannot be measured as much as it is needed (e.g. in space and time). These internal state variables incorporate dynamics of the process, which together with the state transition matrix defines the underlying dynamics of the process (figure A.1).

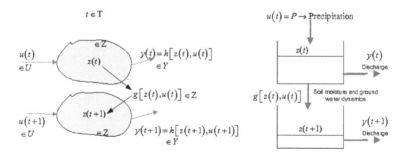

Figure A.1: *Graphical representation of the transition in time of a state space model, the right side represents a hypothetical transition of states based on tanks.*

From the data-driven point of view systems work only with observed input-output data pairs, and do not contemplate explicitly important unobservable variables like internal states, $z(t)$, which are not commonly available, or cannot be measured as much as it is needed (e.g. in space and time). These internal state variables incorporate dynamics of the process, which together with the state transition matrix defines the underlying dynamics of the process.

The equation A.1 and A.2 show a typical representation of state-space model, which is in fact is the discrete-time time-invariant non-linear dynamic state representation.

$$z(t + 1) = g[z(t), u(t)] \tag{A.1}$$

$$y(t) = h[z(t), u(t)] \tag{A.2}$$

where $t \in T$ is time (a set of integers), $z(t) \in Z$, the system state of dimension l, $u(t) \in U$, the input of the dimension r, $y(t) \in Y$, the output of dimension m; $g : ZxU \longrightarrow Z$, the one step ahead state transition function, and $h : ZxU \longrightarrow Y$, the final input-state-output transfer function.

The state transition matrix Φ can be formed by a repeated application of the state equation $g[\bullet]$ in equations A.1 and A.2. If we define $u^* = \{u(t + k + 1), u(t+k)...., u(0)\} \in U^*$, for $k > 0$, then $z(t+k) = \Phi[z(0), u(t+k-1), ..., u(0)]$, where $\Phi[z(0), u(t + k - 1), ...u(0)]$, where $\Phi : ZxU^* \longrightarrow Z$. From A.1 and A.2 we have

$$y(t + k) = h[\Phi[z(0), u(t + k - 1), ..., u(0)], u(t + k)] \tag{A.3}$$

It has been shown by Leontaritis and Billings (1985), that if the system (eq. A.2) can be described by a state-space equation in infinite dimensional space and when the system is close to its equilibrium point it can be approximated by a linear system. For a single-input-single-output (SISO) system, it can be represented in a recursive input-output form as

$$y(t) = f[y(t - 1), ..., y(t - n_y), u(t - 1), ..., u(t - n_u), w] + e(t) \tag{A.4}$$

For $f[\bullet]$ some nonlinear mapping; ny and nu are positive integers representing the lags in the system observable inputs/outputs. In practice, $y(t)$ is subject to noise observations or model mismatch through the noise term $e(t)$ (usually assumed as uncorrelated Gaussian sequence with variance 2).

If this is approximated by a linear system:

$$\widehat{y}(t\,|\mathbf{w}) = -a_1 y(t - 1) ... - a_{n_y} y(t - n_y) + \tag{A.5}$$

$$b_1 u(t - 1) + ... + b_{n_u} u(t - n_u) = \mathbf{X^T}(t)\,\mathbf{w} \tag{A.6}$$

Where $\mathbf{w} = \left[-a_1, ... - a_{n_y}, b_1, ..., b_{n_u}\right]^{\mathbf{T}} \in \Re^n$, $(n = ny+nu)$, is an unknown parameter vector and $\mathbf{x}(t) = [y(t - 1), ..., y(t - n_y), u(t - 1), ..., u(t - n_u), w]^{T}$ is a known input/output observation vector or a set of system repressors. So

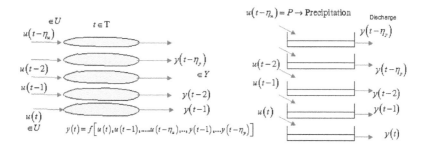

Figure A.2: *Graphical representation of the auto-regressive mathematical model*

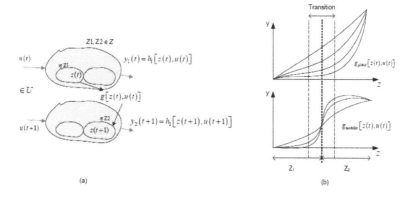

Figure A.3: *Graphical representation of a modular model situation, where due to a change of state-space, significant output is obtained*

the system will become represented only by a linear function that represents the mapping of input-output (fig A.2).

The modular model approach can be visualized as a composite situation were it is assumed that the mapping function $g[\bullet]$, will generate $z(t)$ states in regions with different dominant process. In figure A.2 the state could change from a region $z_1(t) \in Z1$ to $z_2(t+1) \in Z2$, where $Z1, Z2 \in Z$. When these regions have completely different dominant processes the natural representation of a overall model is difficult (fig A.3a,A.3b). This means y can be better represented by creating models for both states $y_1 \in Y1$ and $y_2 \in Y2$, where $Y1, Y2 \in Y$.

Since each model represents an independent formulation of a state-space, the above transformation can be generalized for both models. However, the transition zone presented in the bottom of the figure(A.3b), shows that if the output of the process is highly sensitive to a change in the states, it would be required a fuzzy transition to link both states in the model, as well to link both results in the input-output model.

DATA-DRIVEN MODELS

The general procedure for developing artificial neural networks (ANN), model tree (MT), support vector machines (SVM) and other data-driven models was already mentioned in Chapter 4, however, the details of their structure were not mentioned. The proposed of this appendix is to cover the principles of some of their formulations.

B.1 Artificial Neural Networks (Multi-layer perceptron)

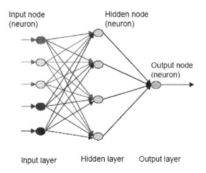

Figure B.1: *Basic diagram of ANN multi-layer perceptron*

The ANN's can be considered an evolution of the regression models. They are essentially mathematical models defining a function $f : X \to Y$. Each type of ANN model corresponds to a class of such functions. The ANN model most used in this thesis was the multi-layer perceptron, explained hereafter.

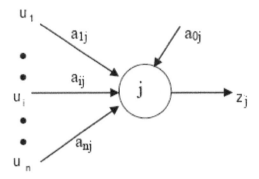

Figure B.2: *Schematic diagram of node j (bottom), Price, 2006*

The inputs of a vector $U = (u_1, ..., u_i, ..., u_n)$ and the corresponding weights leading to the node form a weight vector $A = (a_1, ..., a_i, ..., a_n)$. The hidden node output z, of node j, is obtained by adding all the products and a bias term:

$$z_j = \sum_{i=1}^{n} a_{ij} x_i + a_{oj} \tag{B.1}$$

This linear combination is affected by a non-linear function g. A log function is used to obtain non-linear soft-combination of the results in each node. The solution of a sigmoid function, like the one in Equation B.2 is sensitive the parameter β. The figure of the transfer functions with different values of β are shown in figure B.3. At the end it is important to take into account a bias term that will displace the function and help the fit.

$$g(z) = \frac{1}{1 + e^{\beta z}} \tag{B.2}$$

The transfer function can be also useful as a way to have some extrapolation capacity. The function never reaches the extremes and therefore it leaves a range for unknown ranges of inputs. Some authors add to this and reduce the range of the function by making normalization between 0.1 and 0.9 and therefore it is increase the difference between the limits of the function and the inputs.

The final output is based on the product of the different hidden nodes outputs multiplied by the connection weights (B) to the final output nodes. In the case of rainfall-runoff models this output is only one (discharge).

$$Y_{ANN} = \sum Bg(Z) + B_0 \tag{B.3}$$

The challenge in this ANN model representation is normally the identification of the right weights that will make the function fit the measurement

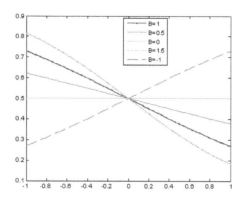

Figure B.3: *Different behaviors of the sigmoid function with different values of β*

values. This has been analyzed by many researchers and therefore different approaches can be found. The most commonly applied is the gradient descent method since its procedure is clearly interpreted with the climbing of a mountain. However, recently simplified methods that approaches the same type of solution in a faster way are being used (i.e. Levenberg Marquadt). This choice is simple a matter of experimental trial and error, and for complex phenomenon it is recommended the gradient descent method (GDM) or other global optimization methods like genetic algorithms (Tsymbal et al., 2005) or adaptive clustering (Solomatine, 1999, ACCO).

To obtain the mentioned weights of the ANN model an objective function need to be defined. The mean square error (Equation B.4) is well known due to its properties in giving higher error weight to the high difference between observed and measured. A solution to this function can be approached by minimizing the MSE.

$$MSE = \frac{SSE}{n},\qquad(B.4)$$

$$SSE = \sum (y_{ANN} - y)^2\qquad(B.5)$$

The ANN weight identification is called training. This starts by replacing y_{ANN} in the equation B.5 from equation B.3, and equalizing it to zero.

$$0 = \sum\sum Bg(Z) + B_0 - Y_{measured})^2\qquad(B.6)$$

The equation B.6 is optimized knowing a number of data samples that are normally measurement from a real life problem. The unknowns in the problem are normally only the weights of the system. The other terms of the function are defined by other processes that help to structure the ANN model (cross validation for nodes identification or validating fit of the optimization procedure).

B.2   Model Trees (M5P)

Following a modular approach to modeling, a data-driven model should be comprised of several sub-models. To train them, the training set may be split into subsets corresponding to a particular sub-process to be modeled, and then each module (Figure B.4) is trained on these non-intersecting subsets(actually, these subsets can be intersecting leading to some versions of ensemble models, but this option is not considered here). When a new input vector is presented, it is first classified to one of the regions (corresponding to the subsets) for which the modules were trained, and then only one module is run to produce the prediction. A class of such methods employing consecutive progressive splits; is typically referred to as trees; examples are: decision trees, regression trees (Breiman et al., 1984), MARS (Friedman, 1991), M5 model trees (Quinlan, 1996).

Since for each data instance (input vector) only one local model is used for prediction, there is a problem of compatibility at the boundary between the regions for which the modules are responsible: for the two neighboring input vectors the predicted outputs could be very different. A solution could be in updating the local models to make them compatible at the boundaries, like it is done in M5 model tree algorithm through smoothing. Wang and Witten (1997) presented M5 algorithm based on the original M5 algorithm but able to deal with enumerated attributes, to treat missing values and using a different splitting termination condition. Several advantages of using the model tree are that it is a non black-box model, understandable, easy to use and to learn, fast in training, robust when dealing with missing data, able to handle large number of features and able to tackle tasks with very high dimensionality. The main procedure of building M5 model trees is as follows:

1. **Building the initial tree** An approach used in M5 trees is to minimize the intra-subset variation in the output values down each branch. In each node, the standard deviation of the output values for the instances reaching a node is taken as a measure of the error of this node and calculating the expected reduction in error as a result of testing each attribute and possible split values. Such split attribute together with the split value that maximize the expected error reduction are chosen for each node. The standard deviation reduction (SDR) is calculated by

$$SDR = sd(T) - \sum_i \frac{|T_i|}{|T|} \times sd(T_i) \tag{B.7}$$

where T is the set of instances that reach the node and $T_1, T_2,$ are the sets that result from splitting the node according to the chosen attribute. The

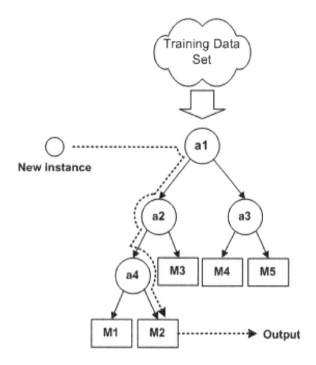

Figure B.4: *Training M5 model trees and their operation. a1..a3 are data partitioning rules; M1...M5 are multiple linear regression models, (Solomatine and Dulal, 2003)*

splitting process will terminate if the output values of all the instances that reach the node vary only slightly, or only a few instances remain.

2. **Pruning** The pruning procedure based on subtree replacement is used in M5 model trees to avoid overfitting. It makes use of an estimate of the expected error that will be experienced at each node for test data. First, the absolute difference between the predicted value and the actual class value is averaged for each of the training instances that reach that node. This average will underestimate the expected error for unseen cases, of course and to compensate, it is multiplied by the factor

$$\frac{n + v * pf}{n - v} \tag{B.8}$$

where n is the number of training instances that reach that node, v is the number of parameters in the model that represents the class value at that node, and pf is pruning factor. The resulting linear model is simplified by dropping terms to minimize the estimated error calculated using the above multiplication factor, which may be enough to offset the inevitable increase in average error over the training instances. Terms are dropped one by one, greedily, so long as the error estimate decreases. Once a linear model is in place for each interior node, the tree is pruned back from the leaves, as long as the expected estimated error decreases.

3. **Smoothing** A smoothing process is used to compensate for the sharp discontinuities that will inevitably occur between adjacent linear models at the leaves of the pruned trees. This is a particular problem for models constructed from a small number of training instances. Smoothing can be accomplished by producing linear models for each internal node, as well as for the leaves, during the time the trees is built. Once the leaf model has been used to obtain the raw predicted value for a test instance, that value will be filtered along the path back to the root, smoothing it at each node by combining it with the value predicted by the linear model for that node. An appropriate smoothing formulation is

$$p' = \frac{np + kq}{n + k} \tag{B.9}$$

where p' is the prediction passed up to the next higher node; p is the prediction passed to this node from below; q is the value predicted by the model at this node; n is the number of training instances that reach the node below; k is constant (in WEKA software we used default value is 15). Smoothing has most effect on a case when the models along the path predict very different values and when some models were constructed from few training instances.

B.3   Support Vector Machines

Support Vector Machines (SVM) is a well known algorithm in machine learning for classification problems. It has been adapted to regression problems showing highly accurate results. Its algorithm aims at minimizing a bound on the generalization error of a model, rather than minimizing the mean square error over the data set. The SVM approach treats the input data as two set of vectors in an n-dimensional space, an SVM will construct a separating hyperplane in that space. The hyperplane should maximize the margin between the two data sets. In this paper, the basic ideas underling SVM are reviewed and the potential of this method for regression (modelling) problems.

Given a training set of model simulations and measurements pairs $(x_i, y_i), i = 1, ..., l \in R^n$ and $y \in \{1, -1\}^l$, the support vector machines(SVM) (Boser, Guyon, and Vapnik 1992; Cortes and Vapnik 1995) require the solution of the following optimization problem:

$$\min_{w,b,\xi} \frac{1}{2} w^T w + C \sum_{i=1}^{l} \xi_i \tag{B.10}$$

, subject to

$$yi(w^T \phi(x_i) + b) \geq 1 - \xi_i \tag{B.11}$$

$$\xi \geq 0 \tag{B.12}$$

In this equation x_i is a vector that is mapped into a higher dimensional space by a function ϕ. Then SVM finds a linear separating hyperplane with the maximal margin in this higher dimensional space. $C > 0$ is the penalty parameter of the error term. Furthermore, $K(x_i, x_j) \equiv \phi(x_i)^T \phi(x_j)$ is called the kernel function. For the scope of this research some experiments have been done with the following kernels.

Linear :

$$K(x_i, x_j) = x_i^T x_j \tag{B.13}$$

Polynomial:

$$K(x_i, x_j) = (\gamma x_i^T x_j + r)^d, \gamma > 0 \tag{B.14}$$

Radial basin function:

$$K(x_i, x_j) = \exp^{(-\gamma \|x_i - x_j\|^2), \gamma > 0} \tag{B.15}$$

HOURLY FORECAST IN THE MEUSE (DELFT-FEWS INTEGRATION)

As shown in Chapter 6, some of the daily models in the Meuse can be replaced by data-driven models with increasing performance. The travel time of the river is one day, therefore hourly resolution is required in operational forecast. For this an adjusted and integrated HBV models is used to generate the hourly forecast (van Deursen 2004). This model includes a more complete routing scheme, and an autoregressive error corrector model. To be able to manage all the model interactions and the comprehensive data-base, the Delft-FEWS environment is used. The the FEWS environment is defined with operation rules solving operational problems with the highest performance. These rules apply on missing data, interpolation and extrapolation of grid and weather station information.

C.1 Methodology

To evaluate the performance of an ANN model in the operational system, it is important to evaluate the forecast performance of the HBV and the routing (SOBEK) model. Three components of the operational system are analyzed:

- Scenario 1: The RMSE of the routing model for the hourly forecast is calculated. The average error obtained is assumed to be the minimum error contribution of this model, to the forecast. This process is done for the overall river basin (Figure C.1a).

- Scenario 2: The analysis of the accuracy of the HBV model using measured information on an hourly basis. The same as in the routing model, the RMSE is used as reference of the minimum error obtained in the subsequent forecasts. This procedure is done for the river gauge measurements

(a) *Scheme of routing model error analysis (scenario 1)*

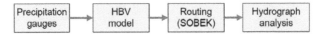

(b) *Scheme of HBV and routing model analysis (scenario 2)*

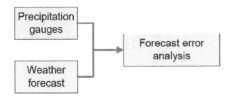

(c) *Scheme analysis of the difference between forecasted precipitation and measured gauges (scenario 3)*

Figure C.1

at Borgharen (overall Meuse) and for the Chooz sub-basin (farest region in the Meuse model).

- Scenario 3: The analysis of the difference between the measured precipitation gauges and the interpolated precipitation forecast.

Note that all the models used in the scenarios include the an ARMA error corrector model (Broersen, 1998), in addition the process-based model (HBV) is combined with an updating procedure to correct the states before each forecast is made. The implication of this is that the starting point of the HBV model results for each forecast is the same as the measured one.

Forecasting model setup

The sub-basin models used for the forecasting vary from the concept given in Chapter 6 only on the way the basins are grouped. The sub-basin grouped can be seen in Figures C.2, C.3, C.4, C.5, C.6 and C.7. This basins have high difference in size and on their seasonal overall discharge contribution. All other information like precipitation measurements and regional hydrological parameters remain the same. Hourly measured data from the river gauges were the basis for grouping the different basins upstream the measurement point.

Hourly measured information from years 2006 and 2007 were used to run different scenarios (Figure C.1). The measured information fed to the model has a number of pre-processes defined in the structure of the forecast modules. The FEWS system is based on XML structure, so a workflow file, was used. The interpolation and routines for filling data, using linear spatial and temporal formulation were used. The weather forecast information is fed into the system with different intervals of 6,12 and 24 hours, depending on the weather agency that generated the information. The HIRLAM and DWD-LM (Germany) weather forecast information were used for the scenario 3.

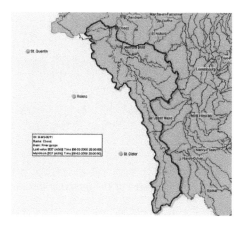

Figure C.2: *Region upstream of the gauging station at Chooz*

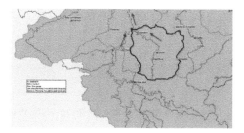

Figure C.3: *Region upstream of the gauging station at Gendron*

C.2   Neural network model (ANN)

The ANN model used for this experiment was a focused time-delay neural network (FTDNN). This is part of a general class of dynamic networks, called focused networks, in which the dynamics appear only at the input layer of a

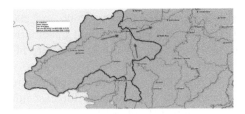

Figure C.4: *Region upstream of the gauging station at Salzinnes*

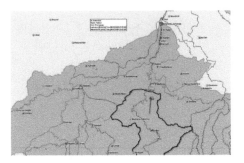

Figure C.5: *Region upstream of the gauging station at Tabreux*

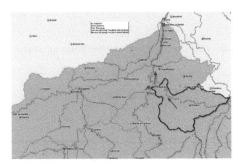

Figure C.6: *Region upstream of the gauging station at Martinrive*

static multilayer feed-forward network. Figure C.8 shows the two layer setup used to build the model.

This neural network concept has the same process as an ANN MLP network when the ANN work only with autoregressive parameters. In this case, the number of lags taken into account was 24 hours for both models developed here (Chooz and the overall Meuse basin at Borgharen). The lags were determined by the autocorrelation of the measured discharge in two years, at both locations.

The total number of hours measured started from November 2005 to December 2007 are 25236. However, the hydrological model had a warming period,

Figure C.7: *Region upstream of the gauging station at Choudfountaine*

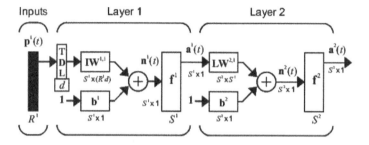

Figure C.8: *Focused time-delay neural network (Mathworks, 2008)*

which was contemplated to be 3012 sample (125 days). The data was split into a training data set of 1.6 years (14016) samples, and verification with the remaining data (e.g. 0.937 years, 8208 samples).

C.3 Results

Scenario 1 and 2 Results of scenario 1 and 2 are shown in Figure C.9, where they are projected in hindcasting (Forecasting assuming the measured values of the input for the models are available, sometimes called perfect precipitation forecast). In the hindcast situation, we could say that we assume perfect precipitation forecast, therefore a constant RMSE line was drawn in Figure C.9. The scenario 1, routing model, the RMSE is 57.23 and for the HBV+SOBEK model is 119. The difference between this two models can be assumed to be contribution from the river basin conception model. In other words, the HBV only carried with less than 50% of the total forecast error at the first time step. The line shown in Figure C.9, shows the forecast performance of the ANN model in terms of RMSE. As we can see, in the first 7 hours the ANN model is less accurate than the routing model with measured inputs (hindcasting). On the other hand, the ANN reaches the accuracy of the HBV+SOBEK model at 31 hours, which is much more than the travel time of the basin. This result shows that a high increase in the performance can be obtained with ANN model, with only past discharge information.

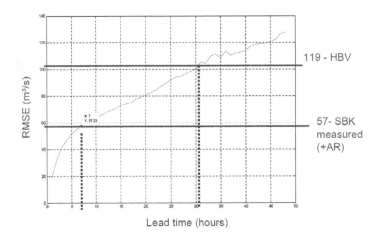

Figure C.9: *Comparison of forecasts made with ANN model, routing model (SO-BEK) and the integrated HBV-SOBEK model (Overall Meuse)*

To extend the analysis, the same analysis was extended to the Chooz region, upper Meuse. In this case, the results are shown without routing component, the HBV only includes the autoregressive error corrector model.

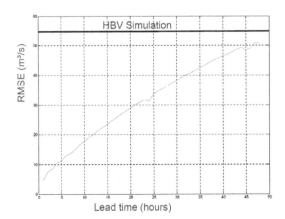

Figure C.10: *Comparison of forecasts made with ANN model and the integrated HBV-SOBEK model (Chooz)*

The Figure C.10 show that the hourly model upstream Chooz with an ANN model autoregressive model could reach 30 hours of forecast lead time with better results than the HBV (RMSE=56 m^3/s).

Scenario 3   An important component in the operational system is the weather forecast information, which is commonly mentioned as the most inaccurate variable in the system. The difference between measured information and the forecast interpolated for the DWD-LM and the HIRLAM for sub-basin one (Lorraine Sud) are presented in Figures C.11, C.12 and C.13. The x axis correspond to the day the forecast was mode, starting 1 of January 2007. The y axis of these figures correspond at the number of hours forecast each day (73 hours). The ranges of precipitation (mm, color) where grouped in 3, as it can bee seen in the right hand side of each figure. The error graph was divided in four ranges due to the inclusion of overestimation and underestimation of precipitation.

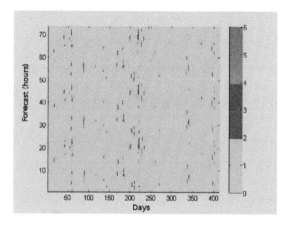

Figure C.11: *Error of interpolated forecast precipitation in hindcasting (gauges measurements) for Lorraine Sud*

On Figure C.12 and C.11 there is no clear indication that the forecast is always worst when the lead time is increasing. On the other hand, in Figure C.13, we can see that most of the error is present in summer (sample 200, around June). This topic was consulted by experts and it seems that this is quite common due to the difficulty to forecast small clouds of rain.

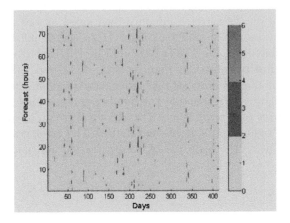

Figure C.12: *Errors of interpolated precipitation forecast using HIRLAM for Lorraine Sud*

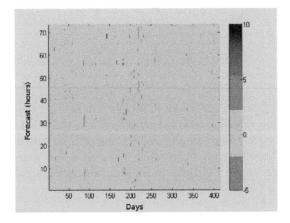

Figure C.13: *Error of interpolated precipitation forecast using DWD-LM for Lorraine Sud*

LIST OF FIGURES

LIST OF TABLES

LIST OF ACRONYMS

ADC	Average discharge contribution
AMI	Average mutual information
ANN	Artificial neural network model
CM	Committee model
CoE	Coefficient of Efficiency
CORINE	Coordination of Information on the Environment (Part of RIZA)
D2P	Hybrid model class, data-driven to physical base
DDM	Data driven model
DPPS	Hybrid model class, data-driven and physical in parallel and serial architectures
DWD	Deutscher Wetterdienst (German Meteorological Survey)
FTDF	First Difference transfer function
FZCM	Fuzzy Committee model
GM	Global data-driven model
HBV	Hydrologiska Byrøans Vattenbalansavdelning model
HBV-M	HBV model of the overall Meuse river basin, results at Borgharen
HBV-S	HBV model of a sub-basin model, results at the outlet of the region in consideration
KNMI	
LC-S	Linear classifier trained for data separation (Part of MM)
LM	Linear regrasion model
M5 and M5P	Model tree
MD	Measured data
MM	Modular Model
MM1	Modular model scheme implemented with clustering
MM2	Modular model scheme implemented with a semi-empirical formulation of baseflow separation
MM3	Modular model scheme implemented with a baseflow fileter formulation
NCEP	National centre for enviromental prediction (USA)
NRMSE	Root relative error, contemplated as Normalized RMSE over standard deviation of the measured.
P2D	Hybrid model class, physical based model to data-driven
PBM	Physical based model
PERS	Persistence index
PRBM	Process based model
RIZA	Rijksinstituut voor Integraal Zoetwaterbeheer en Afvalwaterbehandeling (Institute for Inland Water Management and Waste Water Treatment in The Netherlands)
RMSE	Root mean square error
SMAR	Soil moisture acounting and routing model
TLFN	Time lagged feed forward neural network
WMO	World Meteorological Organisation

SAMENVATTING

Operationeel hydrologisch voorspellen is gebaseerd op uitgebreid gebruik van verschillende typen hydrologische modellen. Het populairst zijn conceptuele modellen, gevolgd door de meer gedetailleerde proces gedistribueerde modellen. Empirische (statistische) modellen worden ook gebruikt, en in het afgelopen decennium heeft deze laatste categorie steeds meer aandacht gekregen vanwege het opkomen van door data aangestuurde modellen. Dit zijn in essentie empirische modellen die gebruik maken van de methoden van machine learning (Computational Intelligence). Een brede keuze aan modellen biedt een bepaalde uitdaging aan een modelleur: deze zal passende modellen moeten selecteren en integreren, en ze vervolgens koppelen aan de gegevensbronnen. Onlangs zijn een aantal studies ingegaan op het probleem om verschillende modelleringsparadigma's te integreren, en het is gebleken dat deze benadering tot een verhoogde nauwkeurigheid van voorspellingen leidt. Meer studies zijn nodig om een consistent modelleringskader te ontwikkelen en dit te testen in diverse situaties. In dit onderzoek worden diverse manieren onderzocht om modellen voor simulatie en voorspelling te integreren. Het stijgende aantal extreme en onverwachte overstromingen in de afgelopen decennia heeft geleid tot een groeiende belangstelling voor nauwkeuriger voorspellings systemen voor hoogwater. Deze systemen zijn noodzakelijk om te waarschuwen tegen overstromingen en daarmee het verlies van levens te voorkomen en schade aan eigendommen en vee te minimaliseren. Daarnaast zijn prognoses van juist lage afvoeren ook belangrijk voor watervoorziening, industrieel gebruik van zoet water, optimalisatie van het functioneren van reservoirs en andere watergerelateerde kwesties. Het doel van de modelleurs is het vergroten van de nauwkeurigheid van de modellen, en de voorspellingstermijn te vergroten. Betere weersvoorspellingen en meer accurate gegevens spelen hierin een hoofdrol, maar ook verbeteringen in modellen en de integratie van verschillende modellen hebben veel potentieel. De keuze voor een bepaald modelleringsparadigma wordt bepaald door de doelstellingen van het modelleren van een hydrologisch verschijnsel, en de beschikbaarheid van gegevens. In het algemeen kunnen de modellen die worden gebruikt voor afvoervoorspellingen worden ondergebracht in drie categorieën: a) fysisch gebaseerde modellen (Physically Based Models - PBMs) (vaak gedistribueerd) gebaseerd op de gedetailleerde weergave van de processen; b) conceptuele modellen en hun meer geavanceerde versie, ook wel proces-gebaseerde modellen (Process Based Models - PRBMs) genoemd, inclusief de zogenaamde "semigedistribueerde" versies; en c) empirisch statistische of gegevensgestuurde modellen (Data Driven Models - DDMs), gebaseerd op de historische gegevens van de gemodelleerde processen. Fysisch-gebaseerde modellen (PBMs) worden in het algemeen gebruikt voor de interpretatie van processen in stroomgebieden. Deze modellen maken gebruik van een groot aantal fysische parameters die op basis van deskundige kennis, veldanalyse, en/of, in complexe situaties, geautomatiseerde calibratietechnieken worden bepaald. PBMs worden vaak gebruikt bij de beoordeling van van overstromingssituaties waar de informatie van de deskundige met de capaciteit van gedetailleerde modellen wordt gecombineerd. Vaak zijn er echter niet genoeg gegevens om PBMs te bouwen, en is voor operationele afvoervoorspellingen een gedetailleerde weergave van een stroomgebied ook niet nodig. Dat is waarom voor 'real time' (operationele) afvoervoorspellingssystemen vaak de voorkeur wordt gegeven aan conceptuele methoden zoals PRBMs en gegevensgeoriënteerde technieken zoals DDMs.

De PRBMs en DDMs hebben een verschillende basis: bij PRBMs is de structuur gebaseerd op de vereenvoudigde beschrijving van de fysische processen, terwijl DDMs gewoonlijk de set van input-variabelen ten opzichte van output-variabelen weergeven. Er wordt vaak opgewezen dat eigenschappen van de PRBMs ontbreken in de DDMs en vice versa. Door de verschillen in deze twee paradigmas is het integreren van dergelijke modellen een uitdaging. Zowel DDMs als PRBMs zijn algemeen geaccepteerd en onderzocht, en hebben nuttige eigenschappen voor verschillende typen problemen. Bij het maken van de beslissing welk type model het meest geschikt is voor een bepaald doel, moet men de mogelijkheid overwegen om beide modelleringsbenaderingen te integreren. Modellen die verschillende paradigma's combineren worden vaak "hybride" genoemd. Bij een dergelijke hybride aanpak moeten de beste eigenschappen van beide benaderingen worden gehandhaafd: fysische concepten van de hydrologische wetenschap in PRBM en de kracht van het inkapselen van historische gegevens in DDM. Bij hybride modellen zijn de verschillende sub-modellen meestal verantwoordelijk voor het modelleren van bepaalde deelprocessen, waardoor het scheiden van de input-ruimte, gebruikmakend van verschillende fysische concepten en/of wiskundige constructies, en vervolgens de integratie van de output nodig is. Als een stap voorwaarts in afvoersimulatie en -voorspelling verkent dit proefschrift het gebruik van geïntegreerde oplossingen met proces-gebaseerde en gegevensgestuurde modellen. Hiertoe wordt voorgesteld een kader voor hybride modellering te gebruiken, en dat te baseren op het beginsel van modulair (lokaal) modelleren.

De belangrijkste doelstelling van dit onderzoek was te onderzoeken 1) wat de mogelijkheden en de verschillende architecturen zijn voor de integratie van hydrologische kennis en modellen met gegevensgestuurde modellen, voor operationele hydrologische voorspellingen, en 2) om deze te testen op verschillende case-studies. De modellen die uit een dergelijke integratie voortkomen staan bekend als hybride modellen. De volgende specifieke doelstellingen zijn geformuleerd:

1. Verkennen van de verschillende architecturen en de ontwikkeling van een kader voor hybride modellen die gegevensgestuurde en proces-gebaseerde hydrologische modellen combineren, in operationele hydrologische voorspellingen, en specifiek in de context van overstromingsrisico's.

2. Verder verkennen, verbeteren en testen van modulair modelleren voor de bouw van gegevensgestuurde en hybride modellen.

3. Verder verkennen, verbeteren en testen van de procedures voor optimaliseren van de structuur van gegevensgestuurde modellen, met inbegrip van de modellen die bedoeld zijn als aanvullings- en/of correctiemodellen.

4. Evalueer de toepasbaarheid van modulaire modellering voor andere verwante problemen, zoals het neerschalen van weersinformatie voor hydrologische voorspellingen.

Dit onderzoek introduceert en ontwikkelt hybride modelleringsprincipes, gebaseerd op modulaire modellen. Algemene classificatie van hybride modellen en een logisch kader van hybride modellen zijn in dit onderzoek ontwikkeld en op basis van dit kader zijn modulaire modelleringsconcepten ontwikkeld en getest op een aantal case-studies. Drie belangrijke principes van modularisatie van de gebruikte modellen zijn: ruimte, tijd en proces-gebaseerd. De belangrijkste case-studie voor de ruimtelijke analyse is het Maas stroomgebied. Rijkswaterstaat gebruikt het hydrologische modelleringssysteem Hydrologiska Byårns Vattenbalansavdelning (IHMS-HBV). Het Maas stroomgebied vertegenwoordigt 15 deelstroomgebieden, elk gemodelleerd door individuele, geclusterde, conceptuele model componenten, die zijn gekoppeld door middel van een versimpeld 'routing scheme'. Dit model is deel van het operationele hoogwatervoorspellingssysteem dat wordt gebruikt in het door Deltares ontwikkelde Delft/FEWS platform. De modellen in dit systeem worden real-time gevoed met regionale weersvoorspellingen waarin wordt voorzien door het Koninklijk Nederlands Meteorologisch Instituut (KNMI). In dit onderzoek zijn verschillende manieren voor vervanging van conceptuele hydrologische sub-modellen door lokale, op data gebaseerde, modellen (ANNs) geanalyseerd. Dit is gedaan op basis van beschikbare informatie (lokaal gemeten afvoeren), en op een studie van de relatieve bijdrage aan de totale modelfout van elk sub-model van een deelstroomgebied. Het resultaat van een dergelijke model-hybridisatie heeft meerdere voordelen, niet alleen in nauwkeurigheid van het algehele model, maar ook in de vergroting van de voorspellingstermijn. Hier speelt ruimtelijke weersinformatie een belangrijke rol in lage en hoge afvoerverschijnselen.

Experimenten met op tijd en proces-gebaseerde modellen zijn uitgevoerd in verschillende typen stroomgebieden in Azië en Europa. De experimenten laten de voordelen zien van het combineren van gespecialiseerde modellen die gebouwd zijn voor verschillende deelprocessen. Het wordt ook aangetoond dat voor de identificatie van zulke deelprocessen het effectiever is om gebruik te maken van hydrologische concepten en beoordeling en kennis van experts, dan van geautomatiseerde data analyse en cluster technieken (die overigens ook erg nuttig kunnen zijn). Het wordt aangetoond hoe globale optimalisatietechnieken helpen om optimale modelstructuren te genereren. Daarnaast worden de mogelijkheden voor het gebruik van modularisatie in voorspellingen voor meerdere tijdstappen vooruit gepresenteerd, en vergeleken met conventionele ANN modellen.

Een uitgebreide gevoeligheidsanalyse van hoofdzakelijk op data gebaseerde ANN modellen is uitgevoerd in dit onderzoek, samen met de analyse van afhankelijkheid van verschillende gegevensgestuurde modellen van verschillende inputs en steekproefsgewijze initialisaties. Deze experimenten bevestigen dat afvoervoorspellende gegevensgestuurde modellen die voorgaande afvoerwaarden gebruiken, gedomineerd worden door autocorrelatie, waardoor de nauwkeurige kennis van neerslag voor bepaalde waarschuwingstijden minder belangrijk is in de algemene fouten analyse. In het algemeen worden ANN modellen met juist gekozen variabelen niet erg beinvloed door verschillende willekeurige initialisaties van gewicht. Met de juiste selectie van variabelen lijkt het zo te zijn dat de correlatie en "Averga mutual information" (AMI) analyse vergelijkbare resultaten geeft voor alle case-studies in deze dissertatie. Van alle gegevensgestuurde modellen die zijn getest, leverden de ANNs de beste prestaties. Wanneer een ensemble van ANNs wordt gebruikt die op verschillende manieren zijn geinitialiseerd, leidt dit tot nauwkeuriger voorspellingen.

Parallelle en aanvullende hybride modelarchitecturen toonden betere prestaties van het voorspellingsmodel dan de ANN en proces-gebaseerde modellen. Meerdere combinaties van ensembles en foutcorrectie modellen zijn getest. Het gebruik van 'committee' (complementaire) modellen die ANN en HBV modellen toepassen voor het Maas stroomgebied laten bijna dezelfde prestaties zien als een model met een foutcorrectie dat gebouwd is met de informatie van voorgaande fouten en voorspellingen van het model. In de Maas case-studie is de non-lineaire foutcorrectie significant beter dan de lineaire foutcorrectie. De resultaten laten zien dat het toevoegen van de foutcorrectie de nauwkeurigheid van HBV verbetert, voor waarschuwingstijden die groter zijn dan de concentratietijd. Het bleek uit experimenten dat een enkele ANN geen nauwkeurige voorspellingen kan doen voor termijnen groter dan de specifieke concentratietijd van de desbetreffende rivier. Deze experimenten gaan uit van een perfecte neerslagvoorspelling, maar kunnen uitgebreid worden naar echte voorspellingen. Over het algemeen werd aangetoond dat de beperkingen van proces-gebaseerde modellen kunnen worden overwonnen door additionele, fout corrigerende, gegevensgestuurde modellen.

Een andere case-studie bestudeert de neerschaling van informatie van General Circulation Models (globale meteorologische modellen) naar meteorologische informatie op stroomgebiedsniveau. De modulaire modelleringsaanpak (gebaseerd op het clusteren van data en het bouwen van aparte modellen voor elk cluster) brengt verbetering ten opzichte van conventionele statistische en gegevensgestuurde modellen. Dit word ondersteund door een case-studie in Ethiopie en data van NCEP uit de VS. De resultaten laten verbetering zien in de algehele nauwkeurigheid van de gemodelleerde neerslag, maar de resultaten voor temperatuur zijn minder overtuigend. Dit laatste kan worden uitgelegd door het feit dat temperatuur een meer periodieke variabele is dan neerslag, en de relatief langzame transitie tussen hoge en lage waarden maakt het een minder passende variabele voor gebruik in modulaire modellen. Samengevat, presenteert dit onderzoek een kader voor hybride modellering, waarin gegevens-gestuurde en conceptuele, proces-gebaseerde modellen op een gecordineerde manier samenwerken en waarin hun rol en prestaties geoptimaliseerd zijn. Ver-scheidene principes van model hybridisatie en modularisatie (ruimtelijk, in de tijd en proces-gebaseerd) zijn in beschouwing genomen en onderzocht in een aantal case-studies. Voor- en nadelen van verschillende benaderingen voor ver-schillende waarschuwingstijden zijn geëvalueerd en bediscussieerd. Voor de Maas case-studie zijn de ontwikkelde modellen bijgesloten als software compo-nenten in het operationele hydrologische voorspellingssysteem voor het Maas stroomgebied en geimplementeerd op het Delft/FEWS Platform.

Deze dissertatie draagt bij aan hydrologische afvoervoorspelling en de re-sultaten kunnen, hopen we, gebruikt worden bij de bouw van effectievere hoogwater-voorspellingssystemen.

Gerald A. Corzo Perez

ACKNOWLEDGEMENTS

The road has been rather long — not to mention somewhat winding.

Over the years it has been my good fortune to encounter many people who have given me more of their time, companionship, professional and personal help, and above all: patience than was perhaps warranted by my seeming determination to finish this thesis.

I would first of all like to thank my promoter, Professor Dimitri Solomatine. He has not only fed me with ideas, given me the scientific support and supervision that a graduate student can expect from his professor, but he also encouraged me to open mind to exploration of the new research avenues and technologies. He always had a friendly ear and was never short of suggestions of so many challenging things. Without him I would never have made it this far.

An important part of this research was Professor Roland Price, who was my promoter in the first year and brought a number of bright ideas related to modelling principles and philosophy of the work. He also provided a valuable input at later stages of this work through regular discussions and useful suggestions.

Dr. Micha Werner (Deltares) and Dr. Marcel de Wit (RIZA, later Deltares) were bringing always the reality and expertise to the understanding of operational hydrological flow forecasting. Their ideas, research, and enthusiasm form the bedrock on which much of this thesis was built. I feel apologetic towards them over the fact that there was not enough time to develop all the excellent ideas that were put forward during our discussions.

Professor Stefan Uhlenbrook of the Hydrology department at UNESCO-IHE, was overlooking the hydrological side of the study and I am thankful for his constant support and very valuable comments. I could not have wished a more thorough discussion partner and sounding board, with his deep understanding of all the nuances of hydrological modelling.

I am also grateful to Jan Luijendijk, Head of the department of Hydroinformatics and Knowledge Management of UNESCO-IHE, for his continuous support in all aspects and creating wonderful working environment. I also learned a lot from you, Jan.

There are three fellow researchers whom I would specifically like to thank for the support they have given me over the years. Firstly, Hong Li for her friendship and numerous discussions and insightful comments that helped me so much. Secondly, Wilmer Barreto, for his useful comments related to technology that helped the development of many of the models. Thirdly, Durga Lal Shrestha, for his friendship and help in establishing the important link between our researches, complementing many of the experiments presented in this thesis.

I would also like to mention fellow PhD students Carlos Velez, Leonardo Alfonso and Arlex Sanchez for their friendship and useful and enlightening discussions (not necessarily about research...). Andreja Jonoski and Ioanna Popescu, who have been great friends. Judith Kaspersma and Schalk Jan van Andel whom specially helped me with the Dutch translation. List of people to mention has, alas, far too many names on it: dear colleagues, roommates and friends with whom I have worked, talked, and had lunches over these years at UNESCO-IHE, my gratitude goes out to all of you. It worth saying that the Institute was like home for me for all these years, and its staff was like a family.

A special word of gratitude, finally, to all the members of the *Delft-Cluster project* and all its partners (UNESCO-IHE, Deltares, TNO, TU-Delft and RIZA) which funded this research and by doing so gave me the opportunity to conduct this research.

Moving towards more personal acknowledgements, I would like to mentioned also my ex-wife, Martha Castaño, which gave me a wonderful daughter and was a support for me for a long time. Many thanks go towards all my family, specially my brother Fabio, and sisters Paula and Aglaya. Lastly let me thank all my other friends[1].

I am, of course, particularly indebted to my mother Maria Clara Perez Rodriguez, my father Fabio Augusto Corzo Salamanca, and my daughter Geraldine Alexandra Corzo for their unwavering heartfelt support and encouragement on all fronts. They have been always there for me, and without them none of the achievements would have been even remotely possible.

Delft,   *Gerald A. Corzo Perez*
September

---

[1]You know who you are

ABOUT THE AUTHOR

Gerald Augusto Corzo Perez was born in Cúcuta on September 16, 1974. He moved with his family to Sheffield (UK) from 1976 till 1980, when he was eight years old he returned back to Cùcuta Colombia. There he finished his basic education and then moved to Bogotà in 1991 where he studied Civil Engineering at the Escula Colombiana de Ingenierìa. He graduated in 1998 and went to work in the planning department and later on became mayor of the municipality of Florencia (Cauca, Colombia). After this experience he joined the University Francisco de Paula Santander in Cúcuta as a lecturer on numerical methods and differential equations, in the department of Mathematics and Statistics. At the same time he followed the specialization program in Teleinformatics and graduated first in his class in 2003. On the same year he joined the Masters program in Hydroinformatics at UNESCO-IHE Institute for Water Education in Delft. His thesis was devoted to the use of modular and computational intelligent algorithms in modelling rainfall-runoff processes. After the graduation in 2005 he joined the PhD programme of the UNESCO-IHE and the Delft University of Technology. Under the supervision of Professor Dimitri Solomatine, he participated in various projects, the most important of which was "Safety against flooding" funded by the Delft Cluster research programme.

Author email: corzogac@yahoo.es

List of publications

- G. Corzo and D. Solomatine. Knowledge-based modularization and global optimization of artificial neural network models in hydrological forecasting. *Journal Neural Networks*, vol 20:528–536, 2007.

- G. A. Corzo and D. P. Solomatine. Baseflow separation techniques for modular artificial neural networks modelling in flow forecasting. *Hydrological Sciences Journal*, vol 52(3):491–507, 2007.

- G. Corzo, D. Solomatine, Hidayat, M. de Wit, M. Werner, S. Uhlenbrook, and R. Price. Combining semi-distributed process-based and data-driven models in flow simulation: a case study of the meuse river basin. *Hydrology and Earth System Sciences Discussions*, vol 6(1):729–766, 2009.

- G. Corzo and D. P. Solomatine. Numerical flow separation and committees of neural networks in streamflow forecasting. In *Geophysical Research Abstracts*, vol 8:09732, 2006.

- G. Corzo and D.P. Solomatine. Multi-objective optimization of ann hybrid committees based on hydrologic knowledge. In *Geophysical Research Abstracts*, vol 8:09794, 2006.

- G. A. Corzo, A. Jonoski, G. Yimer, Y. Xuan, and D. P. Solomatine. Downscaling global climate models using modular models and fuzzy committees. In Shei-Yui Liong, editor, *8th international conference in hydroinformatics*, Concepcion Chile, 2009. World Scientific Publishing Company.

- G. A. Corzo and D. P. Solomatine. Optimization of base flow separation algorithm for modular data-driven hydrologic models. In *7th International Conference on Hydroinformatics*, Nice, FRANCE, 2006. Research Publishing Services.

- G. A. Corzo and D. P. Solomatine. Optimization of baseflow separation algorithms for modular data-driven hydrological models. In *Geophysical Research Abstracts*, Wien, 2007.

- G.A. Corzo, Y. Xuan, C.A. Martinez, and D. Solomatine. Ensemble of radar and MM5 precipitation forecast models with M5 model trees. In *Hydro informatics in Hydrology, Hydrogeology and Water Resources*. Symposium JS.4 at the Joint IAHS & IAH Convention, Hyderabad, India, September). IAHS Publ. 331., 2009.

- S. van Andel M. K. Akhtar, G. Corzo and A. Jonoski. Ganges river flood forecasting using spatially distributed rainfall from satellite data and artificial neural networks. *Watermill publications*, 2007.

- D. P. Solomatine and G. A. Corzo. Learning hydrologic flow separation algorithm and local ann committee modelling. In *IJCNN '06. International Joint Conference on Neural Networks*, pages 5104 – 5109, Vancouver, 2006.

- M. K. Akhtar, G. A. Corzo, S. J. van Andel and A. Jonoski. River flow forecasting with Artificial Neural Networks using satellite observed precipitation pre-processed with flow length and travel time information: case study of the Ganges river basin Hydrology and Earth System Sciences Discussions, 6(2), pages 3385–3416, 2009

- D. Love, S. Uhlenbrook, G. A. Corzo, S. Twomlow. Rainfall-interception-evaporation-runoff relationships in a semi-arid mesocatchment, northern Limpopo Basin. *Submitted Hydrological Sciences Journal*, 2009.

- G. A. Corzo, D. P. Solomatine, M. Werner, M. de Wit and S. Uhlenbrook. Comparative analysis of artificial neural networks, error corrector and committee models in operational flow forecasting. *Submitted to Environmental software and modelling*, 2009.

Printed and bound by CPI Group (UK) Ltd, Croydon, CR0 4YY